FitzRoy and his Barometers

*To Laura, Matthew and William,
who will never know what it is like to have a 'normal' Dad!*

Also by the author and published by Baros Books:
Aneroid Barometers and their Restoration
Barographs
Bizarre Barometers and Other Unusual Weather Forecasters
Care and Restoration of Barometers

Also published by Baros Books:
Antique Barometers: An Illustrated Survey by Edwin Banfield
The Banfield Family Collection of Barometers by Edwin Banfield
Barometer Makers and Retailers 1660–1900 by Edwin Banfield
Barometers: Aneroid and Barographs by Edwin Banfield
Barometers: Stick or Cistern Tube by Edwin Banfield
Barometers: Wheel or Banjo by Edwin Banfield
The History of the Barometer by W. E. Knowles Middleton
The Italian Influence on English Barometers from 1780 by Edwin Banfield
A Treatise on Meteorological Instruments by Negretti & Zambra

FitzRoy and his Barometers

Philip R. Collins

Baros Books

© Philip R. Collins 2007
First published 2007

Baros Books
5 Victoria Road
Trowbridge
Wiltshire
BA14 7LH
UK

All rights reserved. No part of this publication may be reproduced, stored in a retrieval system, or transmitted in any form or by any means, electronic, mechanical, photocopying, recording or otherwise, without the prior permission of the publisher.

A CIP catalogue record for this book is available from the British Library
ISBN-13: 978-0-948382-14-7

Typeset by:
Ex Libris Press
16A St John's Road
St Helier
Jersey

Printed and bound in Great Britain by:
Cromwell Press
Trowbridge
Wiltshire

Cover images: Rear Admiral FitzRoy (courtesy of K. Woodley) with marine and storm barometers from Negretti and Zamba catalogue c.1880 and a domestic FitzRoy barometer from F. Darton and Co. catalogue c.1905.

Cover design: Geo Ashton

Contents

Preface … vi

Acknowledgements … vii

1 FitzRoy and the *Beagle* … 1

2 FitzRoy and the Meteorological Office … 18

3 FitzRoy Marine or 'Gun' Barometers … 40

4 FitzRoy Storm Barometers … 54

5 FitzRoy's Barometer Legacy … 84

6 Buying a FitzRoy Barometer … 122

Bibliography … 141

Index … 143

Preface

This book arose out of what at first was just an interest in a name on a barometer, but soon became something of an obsession. Having handled many barometers to which Admiral Robert FitzRoy's name was linked, it was natural to want to discover more about the man who most people only know as the captain of the *Beagle*, the ship that carried Charles Darwin on his famous voyage around the world. Although FitzRoy's career has been overshadowed by that of Darwin, there are now several good biographies of FitzRoy which I would recommend to any reader who wishes to study further this most captivating man (see the Bibliography). This book is rather an attempt to look more closely at FitzRoy's contribution to meteorology and especially to clear up some common misconceptions about the barometers associated with his name.

A lot of misunderstanding has arisen over the years about 'Admiral FitzRoy barometers'. One customer reported that the antique dealer from whom she had bought a FitzRoy barometer had informed her that Captain FitzRoy used to carve them during his spare time on the bridge of the *Beagle*. Another was told that his barometer was an earlier model by 'Lieutenant FitzRoy' before he became an Admiral. I hope this small work may help to dispel many of the myths surrounding FitzRoy and his barometers for the benefit of owners or would-be owners of these most fascinating instruments.

Cautionary Note about Mercury
Mercury is an accumulative toxin, like many heavy metals (such as lead), and should be handled with proper caution to avoid spills, vapour inhalation, ingestion or skin contact. It is sensible to consider that a mercury barometer might leak, and carry it in a suitable container that will hold any possible spillage. If handling mercury, use appropriate protective wear, such as latex gloves, and store any surplus mercury in a non-breakable container. Mercury also corrodes or forms amalgams with many metals so avoid contact with metal. Anyone intending to handle mercury should consult the relevant Health and Safety Executive guidance and information.

Acknowledgements

I am indebted to many people who have helped me in the production of this book: the staff, past and present, of the Meteorological Office Library and Archive, the Patent Office, the National Archives, the Hydrographic Office Archives, the Shipwrecked Fishermen and Mariners' Royal Benevolent Society, and, especially, Brian Wead, Barry Cox and Peter Moorman of the Royal National Lifeboat Institution (RNLI), who very kindly assisted me personally and made available records which have not generally been consulted and which have added much to my knowledge of the use of barometers in saving lives. The RNLI has also helped in fuelling my interest in FitzRoy's storm barometers, many of which the Institution still possesses. In the true spirit of FitzRoy, their members continue to save lives at peril on the sea.

I am also grateful to Mick Wood for allowing me access to his unpublished notes on the *Royal Charter* storm of 1859, and to Sue Ashton for once more taking a messy and untidy manuscript and turning it into a readable book.

The author and publisher gratefully acknowledge the permission granted to reproduce the copyright material in this book:

Cover picture of Rear Admiral FitzRoy by kind permission of K. Woodley.

Figure 4.6 from *The Life-boat*, 1 January 1861, p. 368, by kind permission of the RNLI.

Every effort has been made to trace copyright holders and to obtain their permission for the use of copyright material. The publisher apologises for any errors or omissions in the above list and would be grateful if notified of any corrections that should be incorporated in future reprints or editions of this book.

Persons who think the barometer intelligible without some intimate acquaintance, who draw hasty conclusions from insufficient observation, are unlikely to judge correctly in unsettled weather, and then blame the barometer.

Robert FitzRoy

1 FitzRoy and the *Beagle*

Robert FitzRoy appears to have been a highly motivated person even in his youth. He was born on 5 July 1805 at Ampton Hall, Suffolk, into a privileged and aristocratic family. His mother, Lady Frances Anne Stewart, was the daughter of the first Marquis of Londonderry, sister to Lord Castlereagh, and the second wife of Robert FitzRoy's father, Lord Charles, a younger son of the third Duke of Grafton. FitzRoy's uncle William and his great grandfather had notable navy careers. His father became a general in the army, and also served for 25 years as MP for Bury St Edmunds, though he is not recorded as speaking in the House of Commons during all this time. It is unlikely therefore that FitzRoy inherited his dedication, perfectionism and work ethic from his father. FitzRoy described him as 'a farming, gardening, and fox-hunting country gentleman' (Mellersh 1968: 19), who first taught FitzRoy how to use a barometer as a weather glass.

In 1809, the family moved to Wakefield Lodge, Potterspury, which had a large park and, perhaps importantly, a lake to sail on. Visitors to the house are told that in his youth FitzRoy went sailing on the lake in a wash tub. FitzRoy's mother died in 1810, and in 1811, at the age of six, FitzRoy was sent to school at Rottingdean, near Brighton. He transferred to Harrow School for one year in 1817, and then, at the age of 12 and a half, entered the Royal Naval College at Portsmouth as a scholar.

These were the early days of college-taught officers (previously, officers learnt by being on board ship from an early age), so that FitzRoy and his contemporaries were often referred to derogatively as 'college boys'. After a number of postings to various vessels on which he gained practical experience, FitzRoy passed his seamanship in 1824 at the age of 19, the first of 26 in his group. He also gained the college medal for mathematics, for which he obtained full marks, having completed additional methods of solving the set exam problems. Soon afterwards, he was promoted to lieutenant.

The following year, he was appointed to the *Thetis* and sailed the Channel coast, the Mediterranean and South American waters under Captain Phillimore, famous for cutting the rum ration from a half to a quarter of a pint. After three years on the *Thetis*, FitzRoy was appointed flag lieutenant to the commander-in-chief of the South American station

Ganges under Rear Admiral Robert Otway. Otway must have seen the talent in this young officer, for when Captain Pringle Stokes of HMS *Beagle*, on its first surveying voyage in South American waters, died after many days suffering from a self-inflicted pistol shot wound to the head, it was Otway who ruled that FitzRoy should take command instead of Lieutenant Skyring, who had been put forward by Captain King of HMS *Adventurer*, the companion ship to HMS *Beagle*. FitzRoy not only had to take command of what must have been a very depressed crew, but he had to win the respect of experienced officers who may easily have resented his appointment.

Thus, in November 1828, at the age of 23, FitzRoy took over the captaincy of HMS *Beagle*, already well into its surveying work, a challenge that he was more than able and willing to meet. During this voyage, travelling around Cape Horn in stormy weather, the *Beagle* went under water twice, bobbing up like a barrel, but each time taking on more water; if the *Beagle* had gone under a third time, it was believed she would have sunk, no doubt with the loss of all life. FitzRoy blamed himself for this near calamity, having not taken enough notice of the falling barometer, an incident he seems never to have forgotten, and may have had a great influence on his later career.

During the surveying work in Tierra del Fuego, FitzRoy took on board a number of native Fuegians, four of whom he brought back to England in 1830 (see figure 1.1). These were named by the crew 'York Minster', 'Jemmy Button' (traded for a large mother-of-pearl button), a young girl 'Fuegia Basket', and 'Boat Memory', who was made a hostage by FitzRoy after the theft of one of HMS *Beagle*'s boats. This action did not help FitzRoy to locate the stolen boat, which was never found, although various pieces of equipment, ropes, oars and so on, were discovered after a long search.

During these months in Tierra del Fuego, FitzRoy decided to offer the four Fuegians free passage to England and promised to return them in due course (although the Admiralty was not asked and later refused to repatriate them). The Fuegians, FitzRoy thought, could be educated in England and, on their return to Tierra del Fuego, would be able to pass on their knowledge to their fellow tribesmen to the benefit both of the Fuegians and of any Western mariners who might seek shelter or supply in those desolate regions. FitzRoy was young and idealistic; as it later turned out, this attempt to 'educate' the people of Tierra del Fuego and establish a Christian mission there was not to prove a success. However, FitzRoy's promise to return the Fuegians to their homeland led to the second (and more famous) voyage of the *Beagle*.

Figure 1.1 Three of FitzRoy's Fuegians (with their estimated ages). Top: Jemmy Button (14 years) in his native dress (left) and in Western dress (right). Bottom left: York Minster (26 years); bottom right: Fuegia Basket (9 years), both in Western dress (from FitzRoy's *Narrative of the Voyage of HMS Beagle*).

On arrival in Plymouth in 1830, the four Fuegians were vaccinated against smallpox, then quarantined on a farm outside the town. Sadly, Boat Memory caught smallpox and died, but the other three survived and, as may be imagined, became something of a spectacle in England. FitzRoy would have been in demand in London society as a renowned 'explorer', and in 1831 he stood as Tory candidate for Ipswich, but was defeated. The Fuegians were taken into the parish of Walthamstow by the Rev. William Wilson, who arranged for their education and welfare as best he could. The Church Missionary Society began to raise public awareness and funds to send missionaries to Christianise Tierra del Fuego. In the summer of 1831, FitzRoy's now famous Fuegians were given an audience with King William and Queen Adelaide, who gave them presents to take back to their homeland.

On 23 May 1831, FitzRoy wrote to the secretary of the Lords Commissioners of the Admiralty, suggesting that 'the proper season for the return of these Fuegians is now drawing near'. He continued with his reasons for having originally brought them to England:

> By supplying these natives with some animals, seeds, tools, &c., and placing them, with some of their own tribe, on fertile country lying at the east side of Tierra del Fuego, I thought that in a few years, ships might have been enabled to obtain fresh provisions, as well as wood and water, during their passage from the Atlantic to the Pacific Ocean, on a part of the coast which can always be approached with ease and safety.
>
> If their Lordships should so far approve of these ideas as to grant me any assistance in carrying them into execution, I shall feel deeply gratified, and shall exert every means in my power; but should they not be thought worthy of attention and support, I humbly request that their Lordships will grant me twelve months' leave of absence from England, in order to enable me to keep my faith with the natives of Tierra del Fuego, by restoring their countrymen, and by doing them as much good as can be effected by my own very limited means.

FitzRoy enclosed a letter in his support from Captain Philip Parker King, the captain of HMS *Adventurer* and his superior officer during the voyage of HMS *Beagle*. It extols FitzRoy's capabilities, and in it we get a glimpse of FitzRoy's zest for work:

> In April I detached the *Beagle*, and *Adventurer*'s tender, to complete portions of the Strait of Magalhaens that were then imperfect; and

by him [FitzRoy], and under his superintendence and able direction, the Magdalen and Barbara Channels through the Tierra del Fuego were surveyed; a considerable portion of the interior sounds on the western coast was examined; and the discovery of the Otway and Skyring Waters was made, by Commander FitzRoy himself, in the depth of the severe winter of that climate, and on which he was absent from the ship thirty-three days in an open whale-boat ... The difficulties under which this service was performed, from the tempestuous and exposed nature of the coast, the fatigues and privations endured by the officers and crew, as well as the meritorious and cheerful conduct of every individual, which is mainly attributable to the excellent example and unflinching activity of the commander, can only be mentioned by me in terms of the highest approbation.

Despite such high recommendation, the Lords Commissioners turned down FitzRoy's request for a commission back to Tierra del Fuego, but allowed him 12 months' leave of absence. True to his word, FitzRoy made private arrangements. On 8 June, he signed an agreement with a London merchant, John Mawman of Stepney Causeway, to be taken with 'five other persons', being his servant, the missionary Richard Matthews and the three Fuegians, back to Tierra del Fuego on a vessel called the *John*, a 200-ton brig then lying in London Dock, for the sum of £1,000.

However, after the intervention of a 'kind uncle' to whom FitzRoy had mentioned his plan, the Lords Commissioners of the Admiralty offered FitzRoy the chance of commanding HMS *Beagle* for a second voyage to South America to continue surveying work, with free passage for the three Fuegians. Now, with the full backing of the Admiralty, FitzRoy could really prepare for his expedition and threw himself headlong into re-fitting and supplying the *Beagle* for its next journey of discovery.

The *Beagle* was by today's standards small: 90 feet long, 25 feet wide, 9 feet deep, and weighing 235 tons. The ship cost £7,803 when built in 1819 at Woolwich, and was one of a class, known as 'half tide rocks' or 'coffins', that easily sank. But FitzRoy knew the ship, was a most remarkable sailor, and set about re-fitting with his usual vigour. The deck of the *Beagle* was raised by 8 inches and by 12 inches forward, and the ship was fitted with then-new lightning conductors by William Snow Harris, a new rudder designed by Captain Lihou, and a Frazer stove which replaced open fireplaces. They carried 5,000–6,000 cans of Kilner and Moorsom's preserved meat, vegetables and soup, and the Medical Department supplied antiseptics and other items useful for preserving natural history specimens.

The *Beagle* was re-rigged by adding a small mizzen mast, and the poop deck was especially arranged for taking surveying instruments. Part of the task was to establish accurate time points around the globe and for this the *Beagle* carried 22 chronometers (six of which FitzRoy paid for personally as he did for two of the six boats that were carried). Figure 1.2 is a cross-sectional drawing of the *Beagle* in 1832, probably a copy of one made in 1897 from memory by Philip Gidley King, who served as midshipman on the second *Beagle* voyage. It is not to scale or technically accurate – men had much less headroom than is shown – but it gives an indication of where some of the items were stored.

Of the barometers on board, there is sadly little information, although we know from a footnote in FitzRoy's later *Narrative of the Voyage of HMS Beagle* (1839) that while the *Beagle* was at Plymouth in 1831 'an excellent marine barometer, made by Jones (with an iron cistern) was sent by water from the maker's hands.' FitzRoy writes that this instrument 'was suspended in my cabin, with the cistern at the level of the sea (excepting during the first eight months, when it was placed six feet higher), and by it all the barometrical observations recorded [in an accompanying table] were taken or corrected.' FitzRoy notes that this barometer was later 'seriously injured' while it was being conveyed by land from Woolwich to London at the end of the voyage in 1836.

According to W. E. Knowles Middleton, in his book *The History of the Barometer* (1964: 157), J. Newman designed an iron cistern in 1823 for a marine barometer, which had a boxwood top to allow air pressure in. We can assume that the marine barometer by Jones that FitzRoy refers to was of a similar design and was worthy of mention by him because its cistern was different from the normal boxwood cistern marine barometers used at the time. It also illustrates FitzRoy's keen interest in the technical details of his instruments, knowledge of which would be useful in his later career.

Sir Francis Beaufort, head of the Hydrographical Office, sent lengthy instructions to FitzRoy on 11 November 1831, which covered many aspects of the voyage, such as major places to visit, the observation of comets and eclipses, daily magnetic readings, the recording of the height of land and tides, and numerous other useful observations that he should make. FitzRoy was also instructed to 'steadily and accurately keep Meteorological Registers, the barometer should be read to the third place of decimals' and to record the temperature of the air and sea. Beaufort recommended that FitzRoy use a particular form, which he enclosed a printed annex of, including a scale of wind speeds which he had invented, each strength of wind being represented by a number. He also suggested that a concise method should be employed to express the state of the weather in the

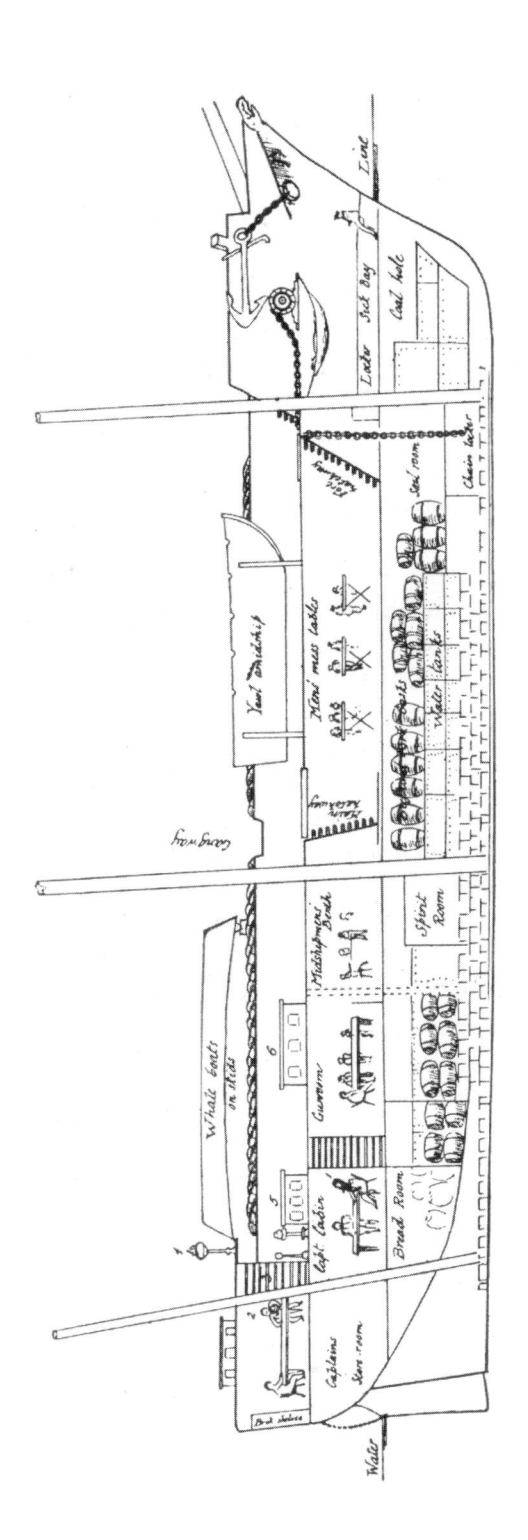

Figure 1.2 HMS *Beagle* in 1832. The numbers indicate (1) Charles Darwin's seat in Captain FitzRoy's cabin; (2) Darwin's seat and (3) specimen drawers in the poop cabin; (4) the azimuth compass; (5) the captain's skylight; and (6) the gunroom skylight (from a 1950 booklet published by Negretti and Zambra).

daily logs. These are hand copied by FitzRoy into the *Beagle* logs as a daily reminder for officers of the watch and are used extensively throughout the log. They are an early form of the Beaufort wind scale we know today:

Figures to denote the force of the wind:
0   Calm.
1   Light Air or just sufficient to give steerage way.
2   Light Breeze or that in which a man-of-war, with all sail set, and clean full, would go in smooth water from 1 to 2 knots.
3   Gentle Breeze or that in which a man-of-war, with all sail set, and clean full, would go in smooth water from 3 to 4 knots.
4   Moderate Breeze or that in which a man-of-war, with all sail set, and clean full, would go in smooth water from 5 to 6 knots.
5   Fresh Breeze or that to which a well-conditioned man-of-war could just carry in chase, full and by Royals, etc.
6   Strong Breeze or that to which a well-conditioned man-of-war could just carry in chase, full and by single-reefed topsails and top-gallant sails.
7   Moderate Gale or that to which a well-conditioned man-of-war could just carry in chase, full and by double-reefed topsails, jib, etc.
8   Fresh Gale or that to which a well-conditioned man-of-war could just carry in chase, full and by treble-reefed topsails, etc.
9   Strong Gale or that to which a well-conditioned man-of-war could just carry in chase, full and by close-reefed topsails and courses.
10 Whole Gale or that with which she could scarcely bear close-reefed main-topsail and reefed fore-sail.
11 Storm or that which would reduce her to storm staysails.
12 Hurricane or that which no canvas could withstand.

Letters to denote the state of the weather:
b   Blue sky, whether clear or hazy atmosphere
c   Clouds, detached passing clouds (cloudy)
d   Drizzling rain
f   Foggy – if thick fog
g   Gloomy, dark weather
h   Hail
l   Lightning
m   Misty, hazy atmosphere
o   Overcast, or the whole sky covered with thick clouds

p   Passing temporary showers
q   Squally
r   Rain, continued rain
s   Snow
t   Thunder
u   Ugly threatening appearances
v   Visible/clear atmosphere, in which objects are distinctly visible
w   Wet dew
.   Under any letter indicates an extraordinary degree

Both forms of notating the wind and weather were used extensively in the record books of the *Beagle*, often six times a day. Barometer readings were also recorded frequently, alongside that of the sympiesometer (see below), but only to two decimal places which was really all that the instruments could be read to.

FitzRoy employed the artist Augustus Earle and George James Stebbing, the eldest son of an instrument-maker, as well as his personal steward, Fuller. He also took Richard Matthews, the missionary, and the three Fuegians, and a young naturalist, Charles Darwin. FitzRoy knew from his previous voyage how lonely the role of captain was, how he and only he had to hold the crew together by strong leadership and personal example on board a small ship in a depressing area of the world where it rained almost constantly, was cold and the land uninviting. It had driven the previous captain to suicide. FitzRoy considered that he needed a companion, someone to dine with and enjoy intellectual discussions.

This was an ideal opportunity for a natural philosopher: there were unknown animals and plants to be recorded, and FitzRoy knew the importance of these discoveries and the use that could be made of them once back in England. FitzRoy and Darwin got on well when they met, and Darwin was later to write that FitzRoy was his 'beau ideal of a captain'. They had disagreements, of course, but discussed many things on the long sea voyage. Darwin never actually went to sea again after the voyage; he was sea sick every day he was on board ship and he was on board for half the length of the voyage, the rest of the time being spent exploring, collecting, observing, and learning.

While we may have a mental picture of press gangs gathering drunken men from the taverns of Plymouth to fill the complement of men needed to sail and run a ship like HMS *Beagle*, this was far from the case. Rather, perhaps, it can be likened to asking for volunteers to go on a space mission today. The *Beagle* was a modern, well-equipped ship, and no expense was spared. If FitzRoy wanted an item, he bought it himself when the Admiralty would not cover the cost. The crew were looking forward to an adventure as well as hard work.

They sailed with 74 people on board. The officers were young, being almost all in their twenties, some as young as 14, including the son of Captain Philip Parker King; most of the crew were under 32. Many were picked by FitzRoy himself, knowing only too well the hardships to be faced and the type of man required. Many had sailed with him on the *Beagle* and *Adventurer* previously. A letter dated 4 July 1831 survives in the National Archives listing the marines FitzRoy wanted on board, which shows that FitzRoy was a thorough man, who knew how to organise the ship and those on board. His first command had taught him much; his next would teach him even more. During the voyage, 25 men deserted and 38 men were discharged (not unusual in those days); there were 611 lashes administered during the five-year voyage, comparatively few by the standards of the day.

For a large part of the voyage the daily routine was well ordered; every man knew his duties, and FitzRoy, by all accounts, led his men well. He kept them busy. In the logs of the *Beagle* we get a glimpse of life on board and how it was methodically recorded. Daily reports include: washed clothes, scrubbed hammocks, aired bedding, employed mending clothes, mustered by divisions, struck the foreyard to overhaul it thoroughly, swayed up the foreyard, employed variously on ship's duty, and on Sundays performed divine service. The daily movements of the ship were minutely and methodically logged, and visitors, unusual events and special occasions were noted, such as:

Saturday 5 April 1834: 'Rec'd on board Jose Maria Luna Arisoner.'
Sunday 6 April 1834: 'East End of Long Island found body of Lieut. Clive (Late of the *Challenger*) lying on beach at high water mark.'
Monday 7 April 1834: 'Capt & officers went on shore & interred the body of Lieut. Clive.'
Friday 27 June 1834: 'Departed this life Mr George Rowlett, purser.'
Monday 13 August 1835: 'Dressed the ship with flags in honour of Her Majesty's Birthday.'

Ships of the day were not easy to maintain, of course; repairs and water proofing were constantly being carried out. On 8 October 1834, for instance, when moored in Valparaiso Bay, they received from the shore a large quantity of carpenter's stores, including 150lbs of white paint, 70lbs of iron nails, 12lbs of copper nails, 50 gallons of oil, 36 sheets of tin, 30 sheets of copper, as well as various files, planes, gimlets, hinges, whitewash brushes, and 39 gallons of coal tar. The sheer scale of supplies to the *Beagle* is staggering. In a 13-day period in August 1835, they received on board a total of 861lbs of fresh beef and about 436lbs of vegetables, so that when in port the men could expect just under a 1lb of fresh beef each and half a pound of vegetables a day, plus whatever else was available, such as bread and biscuits. Their water usage was calculated as 68 gallons a day, with a daily quarter of a pint rum ration. During some of the time, provision must have run short, but they were generally well vitalled. During their surveying work on land, they would hunt and catch fish whenever they could. Occasionally, they would receive provisions from another ship. For example, on Monday 24 August 1835, they received 150lbs of fresh beef and vegetables, 1,994lbs of bread, 600lbs of flour, and a 74-gallon cask of rum from HMS *Blonde*.

During the voyage, the industrious FitzRoy and his officers produced 82 coastal sheets, 80 harbour plans, and 40 views covering southern portions of the South American continent. The quality of the surveying work was exemplary; many of the charts were used for over a hundred years and several were still in use in 1974 when they were superseded by new charts or new metricated editions.

We know that FitzRoy kept records of air pressure from a barometer (figure 1.3), and that a sympiesometer (figure 1.4) was also consulted, perhaps as a comparison. The sympiesometer had been invented in 1818 by Alexander Adie and was considered in some quarters to be excellent for use on board ship as the liquid in it, being an oil, did not move up and down with the erratic movement of the ship like the mercury in a stick barometer. By comparing the readings recorded on the voyage, it is apparent that the sympiesometer was not following the barometer in a linear fashion as the differences in readings vary by differing amounts, sometimes by several decimal points.

A sympiesometer is not as scientifically accurate as a mercury barometer, but it would have nevertheless played an important part in weather predicting. As it is sensitive to temperature (the gas in the reservoir expands and contracts according to temperature, thus altering the level of the oil), the sympiesometer has the appearance of being more sensitive than other barometers in some circumstances. During the first voyage of the *Adventurer* and the *Beagle*, Captain Philip Parker King wrote in his diary for 25 March 1830: 'The sympiesometer was my constant companion: I preferred it to a barometer, as being much more portable and quicker in its motions.' Captain King had some reservations about the barometer. When describing an incident when the barometer failed to indicate in advance a change in the weather, he wrote: 'With respect to

Figure 1.3 A typical mahogany marine barometer of the type that would have been carried on board the *Beagle*. This instrument was made by Thomas Jones, 62 Charing Cross, London, c.1820.

Figure 1.4 A typical sympiesometer of the period in a mahogany case by Alexander Adie, Edinburgh, c.1825.

the utility of the barometer as an indicator of the weather that is experienced of Cape Horn, I do not think it can be considered so unfailing a guide as it is in the lower or middle latitudes. Captain Fitz-Roy, however, has a better opinion of the indications shewn by this valuable instrument ....' It seems therefore that, even in his younger days, FitzRoy was already known for his trust in the barometer.

FitzRoy's journal during the second voyage records the everyday changes and adjustments needed to several barometers and sympiesometers in the daily routine of taking measurements; for example, changing the sympiesometer on 12 May 1834, and adjusting its index four-tenths higher on 20 November 1834 and five-tenths higher on 25 March 1835. There are gaps of several weeks in barometer readings until 10 May 1835 when the deck barometer was used instead of the cabin barometer, and, on 7 June, a sympiesometer was sent onboard the schooner. On 8 June, the barometer tube was discovered to be loose and not used further; and on 13 June a new sympiesometer was put into service. After a week of no recorded barometer readings, the cabin barometer was again used for readings on 30 June, while the sympiesometer needed to be changed on 12 August, and again on 10 September.

It is therefore evident that several spare barometers and sympiesometers were taken and needed to be used. The carriage of delicate barometers on such a small ship would have been very difficult, a fact that FitzRoy would try to remedy in his later career. No doubt this rather unsatisfactory experience with sympiesometers explains why they were not used by FitzRoy in later years, despite their singular advantage over mercury of being more transportable. Then, as now, they were not reliable as scientific instruments, but they remain items of historic interest and curiosity for the collector. Barometers were also taken ashore for observations. FitzRoy took two mountain barometers on shore at the River Santa Cruz during explorations and they were suspended close to the sea and compared with a barometer on board ship which had its cistern level with the sea.

Throughout the voyage, FitzRoy was often challenged by extremes of weather and other natural events which were duly noted in the *Beagle* journal, for example:

22 July 1832: 'Thunder and lightning early in the morning'
20 January 1835: 'eruption of Osorno'
20 February 1835: 'felt severe shock of an earthquake'
16 March 1836: 'passed through a remarkable tide ripple, or meeting of waters'

The *Beagle* was frequently exposed to lightning. As the ship had been fitted with William Snow Harris's new lightning conductors, this was of course of great interest. The lightning conductors seem to have performed to FitzRoy's satisfaction. He reports the ship as having been struck by lightning 'on at least two occasions, when – at the instant of a vivid flash of lightning, accompanied by a crashing peal of thunder – a hissing sound was heard on the masts; and a strange, though very slightly tremulous, motion in the ship indicated that something unusual had happened.'

FitzRoy's *Narrative* is peppered with observations of the weather during the voyage and his own explanations of the type of weather to be expected following a certain appearance of the sky or according to the direction of the wind, at different places and in different seasons. FitzRoy learnt from his observation of the barometer which, he noted, 'stands high with easterly, and comparatively low with westerly winds, on an average. Northerly winds in the northern hemisphere affect the barometer, like southerly winds in the southern hemisphere.' Of a heavy storm during the first voyage of the *Beagle*, FitzRoy commented that 'the barometer foretold it very well, falling more than I had previously seen.'

An extract from the *Narrative* for 12 April 1830 shows how FitzRoy's experience was gained by routine observation of the barometer (or 'glasses') combined with other features of the weather: 'The glasses had at last been rising; and during the past night and this day, the wind was very strong with much rain. The wind shifted from the northern quarter into the southern, drawing round to the S.E.; which of course, would make the mercury rise higher after being so very low, though the weather might prove extremely bad.' FitzRoy paid close attention to such features as cloud. He noted an unusual appearance on 12 May 1832:

> A cloud like a dense fog-bank approached; and as it drew near, the lower and darker part became arched, and rose rapidly, while under it was a white glare, which looked very suspicious. Sail was immediately reduced – we expected a violent squall; but the cloud dispersed suddenly, and only a common fresh breeze came from the foreboding quarter. Neither the sympiesometer nor the barometer had altered at all; but the cloud was so threatening that I put no trust in their indications, not being then so firm a believer in their prophetic movements as I am at present ...

Clouds were first classified into types by Luke Howard. FitzRoy was not only aware of the different types of clouds, but interested in studying them further. In the Appendix to his *Narrative*, he illustrates, in his own hand, 16 types of clouds and describes their forms and classes. In his journal for June 1833, he gives the following account of the appearance of the sky and clouds as predictors of coming weather:

> I have always found that a high dawn ... and a very red sky, foretold wind – usually a gale; that a low dawn and pale sun-rise indicated fine weather; that the sun setting behind a bank of clouds, with a yellow look, was a sign of wind, if not rain, and that the sun setting in a clear horizon, glowing with red, was an unfailing indication of a coming fine day ... soft clouds – clouds which have a watery rather than oily look – are signs of rain; and if ragged, or streaky, of wind also. Light foggy clouds, rising early, often called 'the pride of the morning,' are certain forerunners of a fine day.

FitzRoy also commented on the weather in letters sent back from the *Beagle* to Francis Beaufort at the Hydrographical Office, remarking, for example, on 30 November 1833: 'South East gales are foretold by a high barometer, with cloudy threatening weather and lightning:– a red sky at sunrise, and a high river with strong current running up.' FitzRoy's future was perhaps even then being steered towards the Meteorological Office.

Without authority from the Admiralty, and knowing that he might have to cover the cost himself, with only the hope of being supported in his decision, FitzRoy signed a hire agreement on 11 July 1832 for two schooner rigged vessels, *La Paz* (15 tons) and *La Liebre* (9 tons) in order to proceed more quickly with the surveying work, especially along the narrow inlets and waterways where the *Beagle* would not easily, if at all, travel. By the time the vessels were handed back, the bill was £1,680 for 12 months' hire. Towards the end of the voyage, on 1 September 1835, FitzRoy purchased the schooner *Constitution* (35 tons) with her masts, rigging and gear and one boat for £400 with instructions to the small crew he left on board, under the command of A. B. Usborne, Masters Assistant, to spend approximately nine months completing the survey of the coast of Peru, then to sell their ship and find their way back to England, which they successfully did. The *Beagle* supplied the *Constitution* with provisions and promptly sailed away on the homeward journey. As well as bread, fresh beef and vegetables, the *Constitution* was provisioned with salt beef, salt pork, flour, suet, raisins, sugar, tea, chocolate, preserved meat and vegetables, oatmeal, peas, rum, vinegar and pickles.

The *Beagle* moved on to the Galapagos Islands to restock and prepare to sail westward to complete more time checks using her chronometers. They sailed on to Tahiti in October 1835, and then to New Zealand, arriving at the Bay of Islands on 21 December 1835. At all the places they visited, they recorded new discoveries of animals and plants, as well as the appearance and customs of the local people.

The *Beagle* continued her journey westward to Sydney, Australia, arriving on 12 January 1836, then going south to Hobart Town (Tasmania), into King George Sound and then westward to the Keeling Islands, arriving on 2 April 1836. The ship then sailed to Mauritius, around the Cape of Good Hope, and then continued ever westward to St Helena and Ascension Island, returning to Bahia, on the east coast of South America, on 1 August 1836 to complete the circumnavigation of the globe. The *Beagle* left South America at 10 a.m. on 6 August, heading directly home to Britain via the Cape Verde Islands and the Azores, reaching Falmouth on 2 October and finally docking at Woolwich on 3 November 1836, thus completing an epic voyage that had lasted nearly five years. FitzRoy returned home almost a national hero, and on 27 December 1836, less than three months after reaching Falmouth, married Mary Henrietta O'Brien, who was probably waiting for his return, though we know little about his private life.

In 1837, FitzRoy received the Royal Geographical Society's Gold Medal for his surveying work, and in 1839 he published his *Narrative of the Voyage of HMS Beagle*. In 1841 he was elected as MP for Durham, and in 1842 was selected to attend Archduke Frederick of Austria on a tour of Great Britain; later that year, he was appointed acting conservator of the Mersey. In 1843, he introduced a bill in parliament to establish Mercantile Marine Boards and enforce the examination of Masters and Mates in the Merchant Service; much of today's acts are taken from the same bill.

In April 1843, FitzRoy was appointed Governor of New Zealand, a new British Colony far from the seat of government in London, virtually unfunded and with little defence. Many British settlers were making claims for land and treating the native Maoris with little respect for their rights and customs. The settlers hoped for a governor who would champion their cause in disputes over land, but FitzRoy showed sympathy for the Maoris which was not well received by the settlers. He genuinely attempted to make improvements, working extremely hard as was his way, but without support from Britain and with long delays in communications to and from his superiors at home in Westminster. Three years later, following political changes back in Britain, FitzRoy was recalled amid dissatisfaction among the settlers. As a passenger on the return sea voyage from New Zealand, FitzRoy convinced the officers in charge of the ship to disobey orders during the early hours and drop strong anchors, due to his observations of the private barometers he had on board. He was thus able to save the ship from near certain wreck as the captain took little heed of the falling barometer.

In 1849, FitzRoy was appointed Captain of the *Arrogant*, the first screw-driven navy ship. In 1850, he yielded to the effects of fatigue and anxiety about home affairs, though he soon recovered after absolute rest and a change of air. In the same year, he became managing director of the General Screw Steam Shipping Company but did not seek re-election. His wife Mary, by whom he had four surviving children – Emily (b.1837), Robert (b.1839), Fanny (b.1842), and Katharine (b.1845) – died in the spring of 1852, aged 40. Two years later, on 22 April 1854, he married Maria Isabella Smyth, and his fifth child, Laura Maria, was born in 1858. He became Rear Admiral in 1857 and was promoted to Vice Admiral in 1863.

The year 1854, however, was probably the most significant for FitzRoy's later career and legacy. That year saw the setting up of the Meteorological Office, and Robert FitzRoy was appointed its first chief.

2 FitzRoy and the Meteorological Office

Robert FitzRoy was not alone in seeking to understand the weather and what causes it to change. Over many decades, observation of the weather had been a continuing pastime in Europe and America, both by clergymen in their country parishes and by urban scientists. The invention in Europe, and later introduction to Britain around 1660, of the mercury barometer was the single most important factor in enabling the study of the air and the observation of patterns between the level of the barometer, indicating air pressure, and changes in the weather.

Understanding – and perhaps even predicting – the weather would have enormous advantages, especially for farmers and mariners. For the farmer, it could save his crop; for the mariner, it could save his life, as well as his ship and cargo. The loss of life around the British coast was constant and unpredictable, and there was a pressing need to begin to understand the weather in the hope that storms might be forewarned. Storms at sea could destroy men of business as easily as sink a ship. Large sums of money were involved, and many merchants lost everything when their investment failed to return. Prudent investors would spread their risk, but losses both of life and property through storm damage focused the minds of those influential men of wealth and learning who were in a position to encourage and support meteorological research.

According to FitzRoy, the first systematic endeavour to collect meteorological observations at sea was begun in 1831 at the Hydrographical Office of the Admiralty, and from time to time surveying ships were ordered to make these observations regularly. It is most likely that FitzRoy was referring to the voyage of the *Beagle*, which was ordered to make such observations. In 1838, a system of meteorological observations on an extensive scale was advocated by the British army officer, Lieutenant-Colonel William Reid when he presented a paper on the law of storms to the British Association for the Advancement of Science; he later published several books on the subject.

As early as 1840, James Glaisher was employed to supervise the Magnetic and Meteorological Department at Greenwich, arranging observations and investigations; Luke Howard had earlier studied and named cloud types, which meant that observations could be standardised. The domestic barometer was increasing in popularity; it was not only a status symbol, but a piece of modern scientific equipment with which people could study the weather for themselves. Books and articles were constantly being written about the elusive science of meteorology. In 1845, European meteorologists gathered in Cambridge for a conference which fuelled further debate, but there was no central organisation to oversee the gathering and exchange of information.

William Reid had earlier persuaded the Colonial Office to begin meteorological observations in some British colonies; he also suggested to his superior, John Fox Burgoyne, that observations could be made by the Royal Engineers' stations overseas, and this began in 1851 under the control of another Royal Engineer officer, Henry James. Burgoyne contacted the United States government, seeking to increase the process of observation, and made proposals for the American and British governments to collaborate. The idea was forwarded to Captain Matthew Fontaine Maury of the US Navy and superintendent of the Washington National Observatory. As someone already involved in collecting meteorological data from US ships since 1842, he saw merit in such a project and went further in recommending an international conference to establish a uniform system of meteorological observations. Burgoyne suggested that the Royal Society was consulted and their sympathetic report was later sent to Maury.

The culmination of all these discussions was a three-week international conference called by Maury, which began on 23 August 1853 in Brussels. Britain's representative, Captain Frederick William Beechey, the superintendent of the Marine Department of the Board of Trade, which had been founded in 1850, signed an agreement (against instructions to commit Britain to anything) for the setting up of an official Meteorological Office. The way was therefore paved for the new department.

With Europe, America and, perhaps reluctantly, Britain engaged in conversation between meteorological observers and men of science, often disagreeing as ever, Britain looked for a person to head the new department. It was perhaps not surprising that, in 1854, when Captain Robert FitzRoy was approached, he leapt at the challenge of becoming the head of the new Meteorological Department of the Board of Trade. His position as 'statist' (or statistician) was to gather information to help further the study and knowledge of meteorology as an aid to navigation. As an experienced sea captain, FitzRoy knew what it was like to be at the mercy of the winds and weather. He soon produced standardised forms for ships to report back weather observations, and made sure that ships were supplied with reliable marine barometers so that they could take accurate readings of air pressure (see chapter 3). In the first report of his new department, in 1855, FitzRoy states that:

> In this new office all the valuable meteorological facts which have been collected by the Admiralty, or that can be obtained elsewhere, will be tabulated and discussed, in addition to the continually accruing and more exact data to be furnished in future ... with the two-fold object in view – of aiding navigators, or making navigation easier, as well as more certain – and amassing a collection of accurate and digested observations for the future use of men of science.

It may be easy to think that FitzRoy was just the figurehead of the new Meteorological Office of the Board of Trade, but in fact he *was* the Meteorological Office. With so few staff, he had to be, and he no doubt wanted to be deeply involved in the work. With an initial staff of only three or four to support his work, the challenge for FitzRoy soon became overwhelming. The brief he was given for the new department was perhaps vague, or at least FitzRoy did not consider that he had firm orders to follow, but could interpret his work as he saw fit. He put his own ideas into action and moulded the office he was in charge of entirely as he wanted, although without sufficient funds.

In his 1857 report, FitzRoy notes that 'about 700 months (equal to 58 years of one log) of good meteorological logs have been received from nearly 100 selected ships in the mercantile navy, and about a third of those have been discussed and tabulated ... There are now more than 180 merchantmen supplied with tested and reliable instruments ...'. FitzRoy was keen to praise the quality of reports by awarding the distinguishing marks A, B, C and D; five captains sent in such exemplary registers that they were presented with a valuable telescope by order of the President of the Board of Trade. By the following year's report, there were more than 600 selected ships in the mercantile marine and many men-of-war sending in observations. FitzRoy comments: 'so large a supply of materials has been already obtained, or is in progress (besides what has been received from other sources), that the discussion and publication of results is now the principal object of anxiety, while continuing a diminished yearly collection.'

By the 1862 report, we see that 800 mercantile ships had been supplied with instruments, and 5,500 months of registrations had been gathered. FitzRoy comments in the same report: 'One of the greatest evils of meteorology hitherto has been the practice of incessantly making observations – without very definite objects in view – with the somewhat vague hope that eventually they might become of value; and the natural consequence has been, voluminous records exceeding the grasp of any genius and industry, however combined in individuals.' It was clear that the observations were coming in faster than they could be processed by FitzRoy's small staff.

FitzRoy's 1862 report gives us a glimpse of arrangements at the Meteorological Office:

> The attendance here is necessarily continuous – between ten and six o'clock daily, for some and from eleven to five for others – of the ten persons employed; only two of whom are yet on the regular establishment of the Board of Trade, namely Mr Pattrickson and Mr Babington, my zealous and able assistants.
>
> Specially scientific duties are taken principally by the latter, whose Cambridge education and aptitude for meteorology have enabled him to render good public service. General management in the office, with financial business, correspondence, and much valuable aid in drawing and calculating, are Mr Pattrickson's particular business.
>
> ... Meteorological telegraphy is satisfactorily attended to by Mr Simmonds and by Mr Symons, who also are assiduously engaged in extracting and reducing various meteorological observations, collected on an extensive scale, therefore needing much time for discussion and preparation for printing.
>
> Mr Harding and his son attend to records, stores correspondence, and translation. Mr Strachan has charge of the instruments and optician's duties, aided by Mr Gaster. Two youths carry out our weather reports, or telegrams, and are otherwise actively employed in searching for papers, extracting and copying.

As head of the department, FitzRoy received a salary of £600, W. Pattrickson, £240, and T. H. Babington, £163 15s; G. H. Simmonds and G. J. Symons both received £109 4s; R. Strachan £161 4s; the two Hardings, father and son, both received £78, as did F. Gaster; with the two messengers each earning £36 8s a year.

Added impetus to meteorological research came from a number of natural disasters. In 1854, when Britain was involved in the Crimean War, a severe storm caused havoc and much loss at Balaclava and many supply ships were sunk. Questions were asked back in Britain: could this storm have been warned in advance? Could the recent invention of the telegraph be used to send storm warnings? There was a new momentum to weather observation, but, as with all bureaucratic matters, action was slow, money was lacking, and support was not given wholeheartedly.

The catalyst that set weather warnings into at least second gear occurred in 1859. The *Royal Charter* was a 300 foot long, iron-built steam ship, running between Australia and Liverpool. Sail was the main power used, but the engine was employed to speed up time lost due to calm periods. Most ships were taking three months to complete the trip, but the owners of the *Royal Charter* promised less than 60 days. The ship left Australia on 26 August 1859 carrying passengers and cargo, including nuggets of gold recorded to have been worth £322,440. This was the height of the Australian gold rush, and many people would have carried gold on their persons. The ship arrived off the south coast of Ireland on 24 October, 58 days from Australia, and on the same evening set off for Liverpool. The weather was snowy and cold in parts of Ireland, but was fair with a light breeze when the *Royal Charter* sailed. In Devon and Cornwall, there was a gale which the captain could not have known about; part of the railway line between Dawlish and Teignmouth, as well as Brighton pier on the south coast, were washed away.

At 1 pm on the 25 October the *Royal Charter* approached Holyhead; the sea was calm but the sky was hazy over the mountains. By 4.39 pm, west of Anglesey, the wind was east and force 6, 24 knots and increasing. The ship could have sought refuge at Holyhead harbour, but if it sailed on it would reach Liverpool within 60 days. The ship had weathered worse storms in its time. After 9 pm the wind backed to north and increased to force 10 with 52 knots. The ship was sailing into the wind, with not much coal on board and so was light and riding high in the water, making the hull and upper decks act as sails. The captain began to worry at 10.30 pm; he turned the ship around and had the wind in its sails. The wind increased to force 12 and the captain ordered the anchors to be dropped rather than be driven onto Anglesey. At 1.30 am the port cable broke and an hour later the starboard cable broke. They decided to cut down the masts, but at 3.30 am the *Royal Charter* struck sand and the ship was forced onto rocks at 5.30 am. Earlier, Able Seaman Joseph Rodgers had swum to the shore and, helped by villagers from Moelfre, a rope was dragged ashore and a bosun's chair rigged up. Delays in using the chair as a result of passenger fear cost more lives. In all, 450 people perished and only 41 men survived. Over the next few weeks, £300,000 worth of gold was recovered.

FitzRoy was, of course, involved in the investigation of such a calamity. Another ship, the *Cumming*, had stood off westward only a few miles away and had suffered no damage. FitzRoy stated that no warning signal from land could have averted the consequences of erroneous management, noting that the *Royal Charter* was equipped with excellent instruments (no doubt referring to barometers) which should have given sufficient warning. He implied that indications were overlooked; not so on the *Cumming*. FitzRoy, a hardened sea captain, knew only too well the value of the messages that a barometer could give to the mariner. The tragedy of the *Royal Charter*, often called the 'Golden Wreck', was to be a major part of FitzRoy's studies over the coming years. He wrote: 'Having a collection of facts such as never accumulated on any former occasion of such a remarkable storm, I propose to treat this one cyclone in the fullest details (in the belief that a better type for illustrating may never occur).'

Figure 2.1 shows part of the synoptic chart or weather map that FitzRoy produced after the wreck of the *Royal Charter*, which is quite similar to forecasts we might see today; figure 2.2 gives the explanatory notes for such a chart. FitzRoy was a pioneer in producing these charts, which became common practice and, until recently, were frequently used on television weather-forecast maps which included isobar lines linking pressure readings to show the highs and lows of air masses. FitzRoy began this work with synoptic charts, although the first published chart, for 31 March 1875, appeared in *The Times* ten years after FitzRoy's death (figure 2.3). The *Royal Charter* storm proved a useful object lesson for the study of weather patterns. Whilst not fully understanding weather as we do today, FitzRoy got to grips with air masses and began to understand cold polar currents and hot equatorial currents which he had been studying since the previous year.

The September 1859 meeting of the British Association for the Advancement of Science, held at Aberdeen under the presidency of Prince Albert, resolved that application should be made to the government for the organisation and trial of a system by which the approach of storms might be telegraphed to distant localities. The idea was not a new one: before 1836, semaphoric telegraph had been suggested in America and Europe for the warning of storms. The sinking of the *Royal Charter*, and another storm on 1 November, pushed the need home further. On 6 February 1861, the first trial storm warnings were made by telegraph at Shields on the north-east coast, but the fleet of vessels ignored it and many were wrecked on the 8th or 9th.

By 1862, a storm-warning system was in place designed by FitzRoy, which survived for many decades. On receipt of a telegraphed warning, cones and/or drums were to be suspended from a fixed yard arm on land which was visible from the harbour and out to sea. An upright cone indicated a gale from the north; an inverted cone, a gale from the south; a drum indicated gales successively. For a heavy gale or storm, a cone and drum where hoisted according to which direction the wind would probably

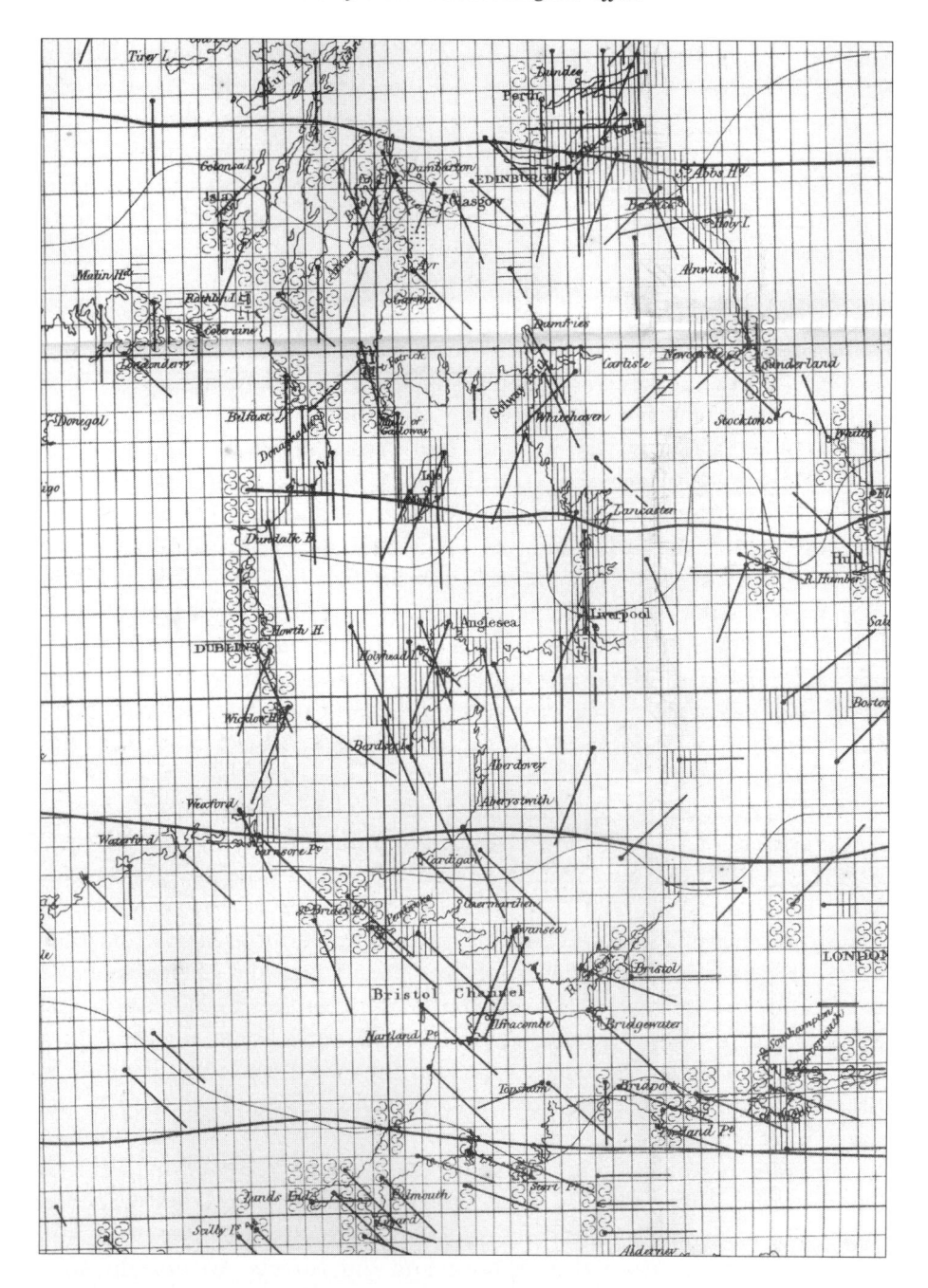

Figure 2.1 Detail of FitzRoy's weather chart of the storm of 1859 that wrecked the *Royal Charter* (from FitzRoy's *The Weather Book*).

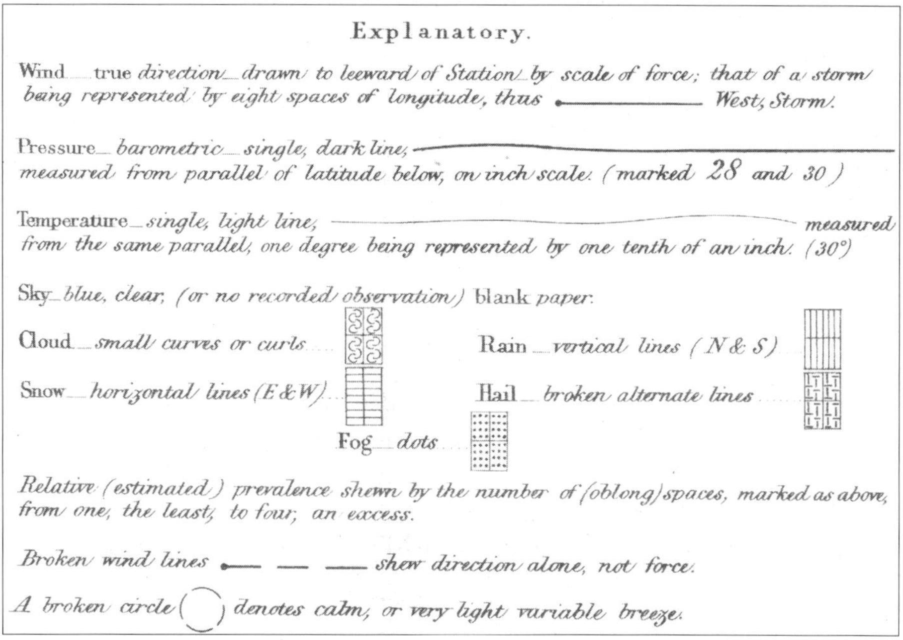

> **Explanatory.**
>
> Wind — true *direction drawn to leeward of Station by scale of force; that of a storm being represented by eight spaces of longitude, thus* ● ———— *West, Storm.*
>
> Pressure — *barometric single, dark line;* ————————
> *measured from parallel of latitude below, on inch scale: (marked 28 and 30)*
>
> Temperature — *single, light line;* ———————————————— *measured from the same parallel; one degree being represented by one tenth of an inch. (30°)*
>
> Sky — *blue, clear, (or no recorded observation)* blank paper.
>
> Cloud — *small curves or curls*
>
> Rain — *vertical lines (N & S)*
>
> Snow — *horizontal lines (E & W)*
>
> Hail — *broken alternate lines*
>
> Fog — *dots*
>
> Relative *(estimated) prevalence shewn by the number of (oblong) spaces, marked as above, from one, the least, to four, an excess.*
>
> Broken wind lines ● — — — *shew direction alone, not force.*
>
> A broken circle ( ) *denotes calm, or very light variable breezes.*

Figure 2.2 Explanatory notes to the weather chart in figure 2.1.

first come from. For night signals, lights were to be suspended in the shape of a triangle or square, as illustrated in FitzRoy's *Weather Book* (figure 2.4).

This simple design, clearly understandable at a considerable distance by fishermen and mariners, is evidence of FitzRoy's empathy with those he was trying to help. Typical of his practical approach to problems, these storm-warning systems survived in some remote areas of the world until after the Second World War, according to people I have spoken to. However, it seems that they were not in use around the British coast during the war: perhaps, as with many other signs, they were removed during the hostilities and never replaced. (It would be interesting to hear from anyone who remembers seeing them in their youth.) Initially, 50 stations were issued with warning symbols by the Board of Trade, but by August 1861, according to a report by Thomas H. Farrer, assistant secretary of the Board of Trade, this had risen to 130. No doubt, as with requests for public barometers (see chapter 3), this number increased as the system gained support from ports and fishing villages.

Thomas Farrer's lengthy report of 12 February 1862 also detailed the number of storm-warning signals that had been issued. There had been a total of 413 signals. Of these, 214 had been followed by a wind that had reached force 8, i.e. fresh gale, or above; the remaining 199 had not reached

WEATHER CHART, MARCH 31, 1875.

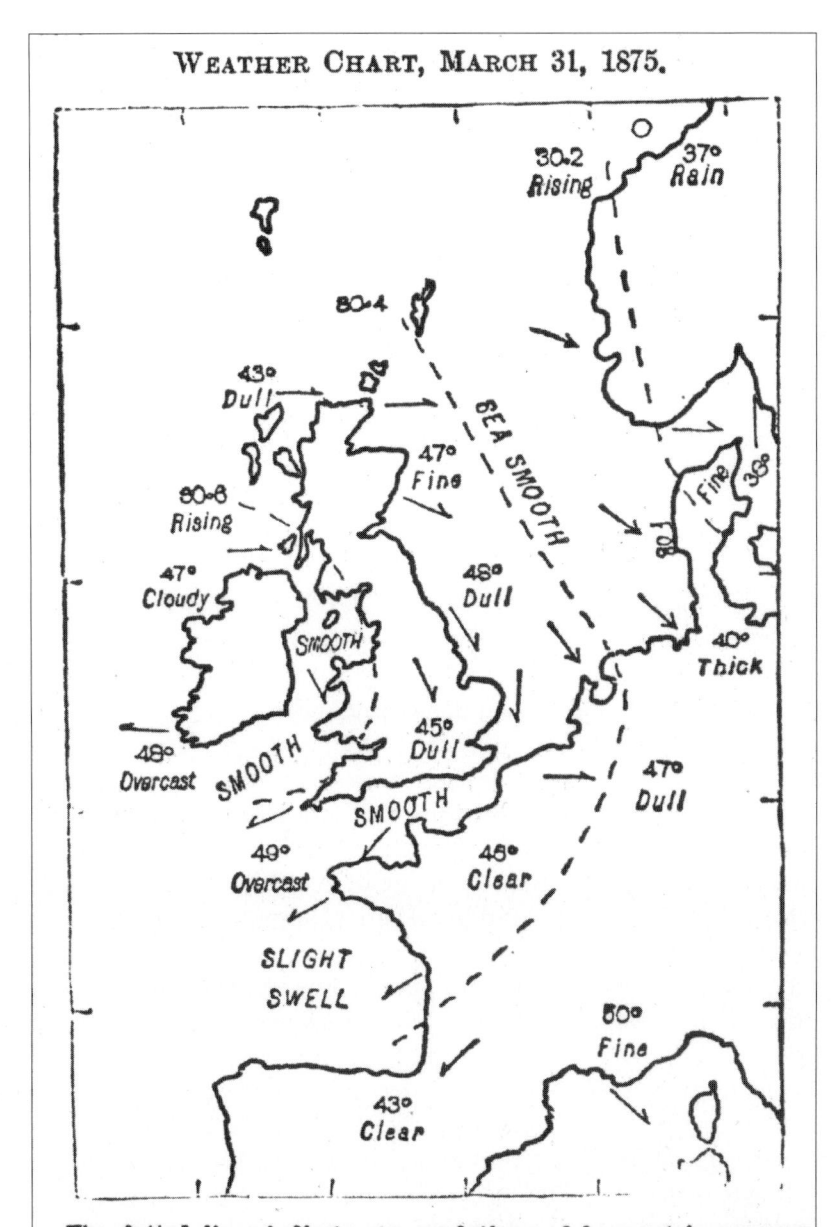

The dotted lines indicate the gradations of barometric pressure. The variations of the temperature are marked by figures, the state of the sea and sky by descriptive words, and the direction of the wind by arrows—barbed and feathered according to its force. ☉ denotes calm.

Figure 2.3 The first published weather chart in *The Times* for 31 March 1875.

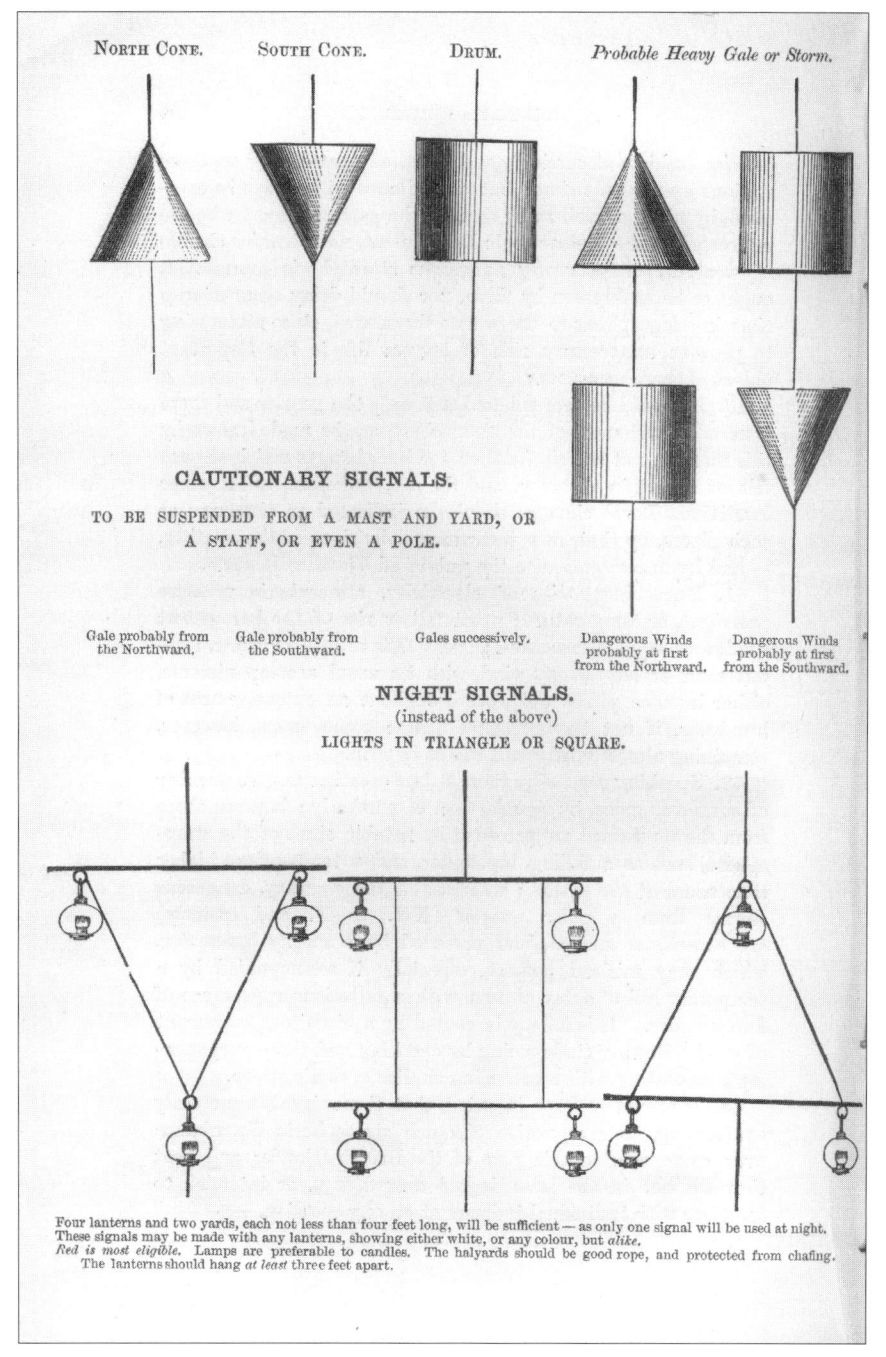

Figure 2.4   FitzRoy's cautionary signals or storm-warning cones (from FitzRoy's *The Weather Book*).

this strength. Of the 214 cases in which the wind had reached gale force, in 72 cases it had reached force 9, i.e. strong gale; in 50 cases, it had reached force 10, i.e. whole gale; in 11 cases it had reached force 11, i.e. storm; and in six cases it had reached force 12, i.e. hurricane. The report, which included further statistics, commented:

> Unless proved to be untrustworthy they show the system to be very far from the perfection necessary to make it of much practical use ...
>
> I hardly know what more to suggest for this purpose myself, unless we were to write to Lloyd's and other bodies interested in shipping, to ask how far the signals had been found to be correct, useful, and trustworthy.
>
> There is no doubt that the public take a lively interest in the matter, and that the signals are popular – and I think very unlikely that anyone will object to the estimate increased as proposed. And I would not stop an experiment likely in the end to be useful.
>
> But the Board of Trade are bound to know what ground they are standing on, what is the exact value of what they are doing; and what is to be the ultimate expense.

Questions were indeed being asked: a letter requesting information as to the usefulness and accuracy of the storm-warning signals was sent out on 4 March 1862 to fishing villages and ports, and replies streamed in. Many supported the continuation of the signals; some had not had time to judge as no or very few warnings had been received; a few were negative; and on one or two occasions it was reported back that boats taking notice of the warning and staying in port were lost when they later put to sea.

FitzRoy answered some of his critics in his 1863 report as follows:

> Many may ask –'Is this system of weather telegraphy sound and advantageous?' – If so why is it opposed?
>
> There are no less than four distinct classes of interested opponents and they should be known. First:– Certain persons who were opposed to the system theoretically at its origin, and having openly expressed, if not published, their objections, are naturally reluctant to adopt other ideas until converted.
>
> Secondly:– A numerous body who cannot have had time and opportunity to look fully into the rationale, but do not realise any want of special information, undervalue the subject, assert it to be a 'burlesque,' and misquote really great authorities.

>   Thirdly:– A small but active party which failed in establishing a daily weather newspaper indirectly opposed to the Board of Trade reports, and have since endeavoured, by conversation, by letters, and by elaborate criticisms in newspapers or periodicals, to exaggerate deficiencies, while ignoring merit in the words of this office, however beneficial their intended objects.
>
>   And fourthly, those pecuniarily interested individuals or bodies who would leave the Coastal and the fishermen to pursue their precarious occupation heedlessly – without regard to risk – lest occasionally a day's demurrage should be caused unnecessarily, or catch of fish missed for the London market.

Whilst it can be argued just how successful FitzRoy and his team were, there is reference to fewer wrecks and loss of life during FitzRoy's time at the Meteorological Office.

Warning of the approach of storms by using the fast telegraph system was one thing, but the idea of predicting the weather a day or two in advance was something that many people at the time would have thought impossible. FitzRoy, however, using the data-collection system of the Meteorological Office and his own considerable experience, felt increasingly able to do this. In fact, he appears to have first coined the word 'forecast' in connection with the weather. Just as the Meteorological Office today forecasts by using a computer-based model that follows certain weather laws, FitzRoy used his knowledge and the information telegraphed to him two or sometimes three times a day, if exceptional changes were happening, to predict the weather.

Queen Victoria, who frequently visited the Isle of Wight, often consulted FitzRoy on the prospects for a calm crossing to the island, and there are numerous mentions of weather forecasts in correspondence between FitzRoy and Sir Emmerson and Lady Tennent. For instance, on 5 July 1864, FitzRoy wrote to Sir Emmerson:

> You only ask what *I think* – a fair question – to which may be replied – *Probably* next Monday 11th will be fine, and hot weather.
>
>   Many *reasons* might be given for this 'forecast' not a mere guess still less a '*prophecy*' but a *probability*.
>
>   The present wind is diminishing and will be less. My belief is that we shall have some summer weather *now* – moderate – fine and warm …

FitzRoy also produced forecasts for the general public, which appeared in *The Times*, and for other establishments, though he still had only a small staff. The first published weather forecast was printed on 1 August 1861, though it arrived with little song and dance, being tagged on to the daily weather report in *The Times* (figure 2.5). It read:

> *General* weather probable during next two days in the –
> North – Moderate westerly wind; fine.
> West – Moderate south-westerly; fine.
> South – Fresh westerly; fine.

A forecast appears to have been drawn up for the day before, but this seems to have been published only later by FitzRoy in the 1862 report of his department.

These forecasts were called 'double forecasts' (two days in advance) and were based on 22 reports received at the Meteorological Office in Parliament Street each morning (except Sundays) and 10 each afternoon, as well as five from the Continent. According to FitzRoy:

> At ten in the morning, telegrams are received in Parliament Street, where they are immediately read and reduced, or corrected, for scale-errors, elevation, and temperature; then written in prepared forms, and copied several times. The first is passed to the Chief of the department, or his Assistant, with all the telegrams, to be studied for the day's forecasts, which are then carefully written on the first paper, and copied quickly for distribution.
>
> At eleven – reports are sent to the Times (for a second edition) to Lloyd's and the Shipping Gazette; to the Board of Trade, Admiralty, Horse Guards and Humane Society. Soon afterward similar reports are sent to other afternoon papers: and late in the day copies, more or less modified in consequence of the telegrams received in the afternoon, are sent out for the next morning's papers.

On 11 April 1862, *The Times*, which waxed and waned in its support for FitzRoy's forecasting abilities, ran the following piece:

> The public has not failed to notice, with interest, and, as we much fear, with some wicked amusement, that we now undertake every morning to prophesy the weather for the two days next to come. While disclaiming all credit for the occasional success, we must however demand to be held free of any responsibility for the too

THE WEATHER.

METEOROLOGICAL REPORTS.

Wednesday, July 31, 8 to 9 a.m.	B.	E.	M.	D.	F.	C.	I.	S.
Nairn	29·54	57	56	W.S.W.	6	9	o.	3
Aberdeen	29·60	59	54	S.S.W.	5	1	b.	3
Leith	29·70	61	55	W.	3	5	c.	2
Berwick	29·69	59	55	W.S.W.	4	4	c.	2
Ardrossan	29·73	57	55	W.	5	4	c.	5
Portrush	29·72	57	54	S.W.	2	2	b.	2
Shields	29·80	59	54	W.S.W.	4	5	o.	3
Galway	29·83	65	62	W.	5	4	c.	4
Scarborough	29·86	59	56	W.	3	6	c.	2
Liverpool	29·91	61	56	S.W.	2	8	c.	2
Valentia	29·87	62	60	S.W.	2	5	o.	3
Queenstown	29·88	61	59	W.	3	5	c.	2
Yarmouth	30·05	61	59	W.	5	2	c.	3
London	30·02	62	56	S.W.	3	2	b.	—
Dover	30·04	70	64	S.W.	3	7	o.	2
Portsmouth	30·01	61	59	W.	3	6	o.	2
Portland	30·03	63	59	S.W.	3	2	c.	3
Plymouth	30·00	62	59	W.	5	1	b.	4
Penzance	30·04	61	60	S.W.	2	6	c.	3
Copenhagen	29·94	64	—	W.S.W.	2	6	c.	3
Helder	29·99	63	—	W.S.W.	6	5	c.	3
Brest	30·09	60	—	S.W.	2	6	c.	5
Bayonne	30·13	68	—	—	—	9	m.	5
Lisbon	30·18	70	—	N.N.W.	4	3	b.	2

General weather probable during next two days in the—
North—Moderate westerly wind ; fine.
West—Moderate south-westerly ; fine.
South—Fresh westerly ; fine.

Explanation.

B. Barometer, corrected and reduced to 32° at mean sea level ; each 10 feet of vertical rise causing about one-hundredth of an inch diminution, and each 10° above 32° causing nearly three-hundredths increase. E. Exposed thermometer in shade. M. Moistened bulb (for evaporation and dew-point). D. Direction of wind (true—two points *left* of magnetic). F. Force (1 to 12—estimated). C. Cloud (1 to 9). I. Initials :—b., blue sky ; c., clouds (detached) ; f., fog ; h., hail ; l., lightning ; m., misty (hazy) ; o., overcast (dull) ; r., rain ; s., snow ; t., thunder. S. Sea disturbance (1 to 9).

Figure 2.5   The first published weather forecast from *The Times*, 1 August 1861.

common failures which attend these prognostications. During the last week Nature seems to have taken special pleasure in confounding the conjectures of science.

Weather forecasting has ever since carried the burden of much criticism. These early official weather forecasts were suspended for awhile until more experience was gained, but then resumed until 18 June 1864, after which *The Times* discontinued them.

It may perhaps have been a desire to justify his theories about the weather that prompted FitzRoy to publish a book on the subject, entitled *The Weather Book*. The volume was written for the ordinary person whom FitzRoy may have thought would benefit from a better understanding of the weather and its changing patterns. He began writing the book on 10 August 1862 when he was in Brighton on what he termed a 'so-called holiday', which seems to have been a period of enforced rest, despite frequent visits to the office. He continued writing the book after his return from Brighton, working at home late into the night. It was apparently completed after only four months and a proof copy was available in late December 1862.

FitzRoy had been working on his own weather theory – that of lunar and solar effects on the weather – and decided at the last minute to include a chapter explaining his theories on the effect of the sun and moon on our weather. It was an idea that did not have much support amongst his contemporaries, but as soon as his book was in proof form he sent a copy to Sir John Herschel, a well-respected scientist of the day, in the hope (with some reason) that he would verify FitzRoy's lunisolar theory. Unfortunately, FitzRoy had to wait until 14 March 1863 to receive a letter from Herschel rejecting his lunisolar theory outright.

He had a more positive response in the summer of 1863 when he was asked by E. H. Marié-Davy, the newly appointed Director of Magnetism and Meteorology at the Imperial Observatory in Paris, for permission to translate *The Weather Book* into French, a request that FitzRoy readily agreed to. This new acquaintance produced long exchanges about weather theories, and it seems that after Marié-Davy had received all of FitzRoy's thoughts on weather forecasting – in considerable detail – Marié-Davy began his own storm-warning system in France in August 1863, much assisted by his correspondence with FitzRoy.

The Weather Book, which was published in early 1863 by Longman, Roberts and Green, calls on an incredible amount of work that had been undertaken over many years. It ran to more than 440 pages, including more than 100 appendixes and 16 charts and maps. FitzRoy sold the copyright outright for £200 and the book sold for 15 shillings, going into a second edition as quickly as March 1863. In *The Weather Book*, FitzRoy makes reference to his own experience on board the *Beagle*, especially as it related to storms and weather phenomena that he and his crew had encountered around the world. It also draws together what were then the latest theories about weather patterns, as well as FitzRoy's own thoughts and theories. In the book, for example, he explains why the weather in the British Isles predominantly comes from the south west: as the Earth revolves, the unevenness of the land drags the mass of air around after it, which, for the majority of the time, attempts to move towards the North Pole; thus, in the northern hemisphere, the weather mostly comes from the south west (and conversely for the southern hemisphere).

In *The Weather Book*, FitzRoy also includes some discussion of the barometer, which 'should be placed where it may be seen at any time, in a good light, at the eye level: and it should be set regularly, at least twice a day.' He explains that it requires considerable experience to be able to interpret barometer readings correctly: 'only those who have long watched their indications, and compared them carefully, are really able to conclude more than that the rising glass *usually* foretells less wind or rain, a falling barometer more rain or wind, or both; a high one fine weather, and a low one the contrary.'

However, according to FitzRoy, although these general conclusions are useful, they are sometimes erroneous, and he includes a series of observations –'the results of many years' practice and many persons' experience' – to help those not accustomed to using a barometer to do so successfully. A major point that he emphasises is that judgements about the weather should be formed, not by the current level of the mercury, but whether the mercury is rising or falling, 'and from the movements of immediately preceding days as well as hours, keeping in mind the effects of change of *direction*, and dryness or moisture, as well as alteration of force (or strength) of wind'. FitzRoy is careful to make the point that the state of the air foretells coming weather rather than indicating its present state, and that 'the longer the time between the signs and the change foretold by them, the longer such altered weather will last; and, on the contrary, the less the time between a warning and a change, the shorter will be the continuance of such predicted weather'. This is clearly the origin of his well-known verse:

Long Foretold – Long Last
Short Notice – Soon Past

which is reproduced on many of the 'FitzRoy barometers' described in chapters 4 and 5. FitzRoy also indicates that the barometer needs to be used in conjunction not only with a thermometer, but also with other instruments to gauge the moistness, direction and force of the wind. In addition, the '*appearances* of the sky should be vigilantly noticed'.

A positive review of *The Weather Book* appeared in 1863 in volume 3 of the *Intellectual Observer*. The publication was full of praise for FitzRoy's weather-forecasting methods and storm-warning signals:

> Few men have done so much practical good with so little pretence as Rear-Admiral FitzRoy, whose forecasts of the weather are looked for with eagerness all round our shores ... when Admiral FitzRoy hoists his alarm signals at the ports, a storm is probable, and prudence commands small vessels or weak ones not to tempt the danger of the seas.

The review described *The Weather Book* as:

> a work which will interest and instruct many readers who would be alarmed at the sight of a more formal treatise, and which is illustrated by numerous valuable diagrams, to which the student will be glad to refer. The present edition is a handsome one, and necessarily expensive; but looking to the popular interest of the questions discussed, and the good that would result from placing the FitzRoy philosophy within the reach of seafaring men and agriculturists, we hope the respected firm of Longmans, to which the copyright belongs, will be able to accommodate itself to modern ideas and publish a cheap *Weather Book* without delay. (pp. 103–9)

Although the book was a success and went into a second edition, the copy I acquired some years ago was unread, having, for the most part, not had its pages cut from its original printing. While it was exciting to cut these pages and be the first person to look at the text since it was printed in 1863, it has to be said that much of the book makes for rather dull reading. A certain amount of criticism emerged after its publication that some of FitzRoy's explanations were long winded and difficult to understand. These criticisms appear to have hit home as can be seen in the preface to the second edition:

> In preparing a second edition of this book I have endeavoured to profit by many judicious criticisms, and have altered or re-arranged

several obscure passages. It is, however, impossible to make so complicated a science as the higher Meteorology quite clear and easy to a person not in some degree acquainted with the subject. Professor Dove says, in his preface to his latest work [*The Law of Storms*], 'The problems which are presented to us by the atmosphere are too complicated to allow of their solution off hand.' Even his writings – explicit as they are to some readers – have been thought obscure by others.

Meteorology never can be an *exact* science, like Astronomy, because its elements are incessantly changing, in *nature* as well as quantity; but it does not therefore require a merely superficial degree of attention.

Although FitzRoy did not understand weather patterns as our meteorologists do today, he had perhaps the best understanding of his day: he was aware of air masses and his study of his synoptic charts put him in a better position to understand the gyratory movements of air masses with their troughs and crests. FitzRoy certainly set the foundations for our modern understanding of meteorology.

We are fortunate to have a wealth of documents surviving from the early Meteorological Office, now stored in the National Archives, and are indebted to Derek Barlow for producing a comprehensive index of these papers, which are invaluable for anyone researching the early days of the Meteorological Office. As an example of FitzRoy's involvement in the day-to-day running of his office, as well as his interest in communicating with other people throughout the country and throughout the world, quoted below is just one of the thousands of letters FitzRoy wrote as head of the Meteorological Office. It was written on 23 October 1863 to E. H. Marié-Davy, who started the French storm-warning service:

My Dear Sir, I have just been much gratified by the perusal of Mr Le Verrier's and your letters and their postscript dated yesterday. I shall call on General Sabine presently, and will communicate to him your kind intentions on behalf of our Royal Society in the matter of the 'Bulletin' and your own work on electricity.

There is nothing, as yet in our reports today, from the *west*, to indicate approaching bad weather, or much wind, but the contrary – some duration of fine days. May I be so free as to say that although *between* the *greater* circuits of a *number* of lesser ones – say cyclones (tourbillions) occur – they are only, as it were, eddies between principal streams – or great *breadths* – orbital extensions, or circuitous movements of air reaching from central to polar regions, and *conversely*, while always (in the temperate zones) advancing slowly *toward the east*. Hence, in our forecasts, we should look at the *main* currents in their *breadths*, and advances, as well as to the eddies occasioned by them between their respective preponderances, or their entire possession of a district of land or water. Then there is the lunar influence with its 7-day period, that of the sun with another periodicity – and even a third which I have not yet traced accurately.

These three influences of unequal periods, acting together at times – but usually at more or less in mutual opposition – it is as interesting as it is yet difficult to trace and distinguish sufficiently.

In my 18th chapter [of *The Weather Book*] are some observations on these subjects which hitherto I have not found it desirable to modify – and which I hope soon to extend, in a similar direction, on the same principles.

May I draw your own attention to the enclosed specimen diagram? I find such a record, if faithfully kept twice or even once a day, is to my mind (through the eye) what an 'indicator' of steam with its 'card' is to a marine steam engineer, – or what a stethoscope is to a physician.

As the general body of the atmosphere *has been* affected during some days – *just past* – one may suppose it will have a tendency to *continue*, a proportionate time, unless, or until interrupted by some sudden, perhaps overpowering cause on a *great scale*.

All the atmosphere within a few hundred miles, if uninterrupted by *high ranges* of mountains, seems to *undulate*, or oscillate, like large, extensive, but (comparatively) shallow *waves*: and is distributed or recovers its equilibrium – with normal quiet, more or less gradually – seldom by *sudden* changes – except in connection with a storm. Pray believe me most respectfully and truly yours. R. F.

P.S. I send you another copy of our telegraphic instructions (on account of three slight typo. errors in page 15).

Also, our *actual* telegrams from a few special stations this afternoon (*just arrived*) by which you may see the *latest* state of the western horizon. Pray return these telegrams (only) as they are *original*, and must be filed?

I find the camphor glasses exceedingly useful – indicating *polarity* (or + electricity) or, the absence of it, in the air – either at earth surface – or *far* overhead – a mile, say, or two, even three miles and *I* believe (heterodox as it may appear to many) that active electricity – such as our air, our body, – everything on earth feels, is stimulated, excited or developed by themselves set in motion by heat and cold – by solar action, against the normal cold of space, take in most visible effects in the torrid and polar regions – respectively as *secondaries*.

Notice the so called electric *'wire-currents'* so often simultaneous with the crossing of polar air over tropic currents – (or under them) and never active (so as to interfere with telegraphing) – except before *strong* polar winds – (when you have *aurora*, and meteors – often shooting stars also) I tell you these *ideas* freely – because I am sure you will appreciate them – in seeking for truth – unfettered by preconceived views. R. F. (Letter from FitzRoy, BJ7/793 89249 To E. H. Marié-Davy)

This letter not only indicates FitzRoy's depth of knowledge and level of involvement in the work of the Meteorological Office, but also, written towards the end of his life, perhaps shows, in its somewhat strange jumble of items, an excited and highly active brain. (It was common for FitzRoy to emphasise words by underlining, which is represented here by italic.)

In his earlier days on board the *Beagle*, FitzRoy had once attempted to give up the captaincy during a period of depression, though his officers had persuaded him not to do so. In early 1850, he was 'tired out' and yielded to the effects of fatigue, though he recovered himself after a change of air. FitzRoy worked himself hard: while writing *The Weather Book* he worked into the night, interrupting his night's rest and frequently leaving himself unable to read or write in the evening without falling asleep. His gradual loss of hearing increased, especially when under strain. In December 1863, in a letter to Lady Emmerson Tennent, he described himself as 'so trebly worked that to no one have we gone out – except my wife's Mother'. During April 1865, he was recommended rest at home by his doctors, but he could not rest, although he was tired. He would make odd trips to the office only to return exhausted by exertion that he could easily have coped with a few years before. Figure 2.6, taken shortly before his death, shows him an old man, thin and worried.

On Saturday 29 April 1865, FitzRoy visited the American, Captain Matthew Fontaine Maury, a key player in the meteorological arena for many years, who was concluding a visit to Britain. Captain Maury could be considered to be FitzRoy's counterpart in the US; they appear to have had disagreements, but Maury was in a prime position of influence in a scientific field that was only just emerging and in which there was much discussion over competing theories. On returning home, FitzRoy was elated and full of ideas; he considered visiting Maury again, obviously excited by the discussions he had had. His wife and he retired around midnight

Figure 2.6 Robert FitzRoy, shortly before his death (from a publication by Negretti and Zambra).

and, although complaining of the light from the window, he had a restful but not refreshing night. He rose some time after 7 am, and entered his dressing room. After awhile, he bolted the door and ended his own life.

So ended, prematurely, the life of a highly intelligent, industrious, ambitious, highly principled and amiable man; one who, without any doubt, fashioned the weather forecasting future of the Meteorological Office and introduced systems that saved and continued to save countless lives. In recognition of FitzRoy's contribution to weather forecasting, the government stepped in with an offer of a house for his widow, who was left penniless on FitzRoy's death. (Just why his finances were exhausted is not known, though he always spent his own money freely in the course of public service, as we saw when he was fitting out the *Beagle*.)

When the committee of the Royal National Lifeboat Institution expressed its regrets on FitzRoy's death to his widow, she replied on 11 May 1865:

> My noble husband sacrificed his life far more than the man who loses it on the field of battle, or the deck of a man-of-war, hotly contending with a foreign foe – more even than those brave men of whom England is so justly proud, who man the life-boat to rescue their fellow-creatures – for he continually periled his life; he gave himself, and all he held dearer than life, for his country; he still held fast to his post – clung to the helm as long as life lasted; and when that enthusiastic spirit was all but worn out, his poor mind succumbed.
>
> All that is left for me to say is, 'God's will be done,' however mysterious in its working. May my dearest husband's memory be honoured to the utmost, is the crying wish of his most disconsolate widow.

3 FitzRoy Marine or 'Gun' Barometers

FitzRoy appears not to have lost much time once ensconced in Parliament Street, the new headquarters of the Meteorological Office. He knew the value of obtaining barometer readings from around the world and of compiling data – it was what his new job was all about. His title was 'statist', or statistician, for the Meteorological Office of the Board of Trade. He was to gather information, and later, it was hoped, some natural laws governing the weather would be discovered from the accumulated data.

Ships already carried marine barometers, made by many different makers, and they were well known to be of use in weather prediction. The navy had been using them for many years. Three letters in the archives of the Hydrographical Office (outgoing letters book) from Francis Beaufort, each dated 29 March 1843, highlight an attempt to organise the carrying of barometers in Royal Navy ships, which hitherto had been rather spasmodic. One letter, to a Captain Washington, reads:

> Sir, I wish you might be so good as to look at the Marine Barometers (there are 17 I believe) at Woolwich Dockyard for Steamers – and tell me what sort of instruments they are, and who made them.
>
> Barometers are to be supplied to all ships – which I wish I had known when at Woolwich that I might have asked leave to see them.

Another was sent to the 'Storekeeper General', asking what price had been paid for the barometers in store and who had made them. A third letter reads:

> Sir, Having been commanded by the L.C.A. [Lords Commissioners of the Admiralty] to provide Marine Barometers for a number of HM Ships, I have to desire, if you wish to participate in that supply, that you will, on or before the 10th of April, send samples of those instruments to this office with the prices offered.

It was sent to the following instrument-makers:

Mr Jones – 62 Charing Cross
Mr Simms – Fleet Street
Mr Dollond – St Paul's Church Yard
Mr Cary – Strand
Mr Worthington – Piccadilly
Mr Gilbert – Fenchurch Street
Mr Pastorelli – Cross Street Hatton Garden
Mr Bate – 21 Poultry
Mr Dennis – 110 Bishopsgate Street

Clearly, moves were afoot to issue all ships with barometers, but the ones then in regular use are not recorded and presumably by various makers. Whether FitzRoy had anything to do with this 'command' from the Lords Commissioners of the Admiralty is mere conjecture. Perhaps he had discussions with Beaufort after his famous voyage or lobbied the Lords Commissioners in other ways. This attempt to standardise the issue of barometers, however, seems not to have been successful as, nine years later, after assuming his new post in 1854, FitzRoy clearly set about arranging methodically for the supply of uniform instruments to ships.

The earliest record of a marine instrument (number 59) being issued by FitzRoy is one to the *Lady Hodgkinson* on 27 November 1854. Shortly after, on 14 December 1854, the agencies listed in table 3.1 were supplied with marine barometers in order to be ready to issue them to ships so that they could return the statistics that FitzRoy needed. From these 15, which were all made by Patrick Adie of London, one of the sons of Alexander Adie, the eminent Scottish maker, there followed regular issues of instruments to various ports and then to ships on behalf of the Board of Trade.

In FitzRoy's 1855 report, he states that:

> instruments, charts and books have been placed on board more than 50 merchant ships, and thirty men-of-war ... Many more ships might have been similarly provided with instruments, had the willingness of their Captains alone affected the supply: but as good marine barometers require time for their construction, and cannot be well made except by skilful and practised opticians, the supply of them has not quite equalled the sudden demand. Moreover, only a certain number can be purchased by the Government annually.

I suspect that the latter reason was the more likely explanation for the number of barometers supplied.

Table 3.1 Adie marine barometers supplied on 14 December 1854

Agency	Instrument no.	Ship	Date issued
Glasgow	37	*Victory*	19/3/1855
Glasgow	38	*Spectre*	23/2/1856
Newcastle	39	*Corremulgie*	11/2/1860
Newcastle	40	*Sir R. Abercrombie*	23/9/1857
Liverpool	41	*Cambridge*	22/12/1854
Liverpool	42	*Crest of the Wave*	12/4/1855
Liverpool	43	*William Hutt*	9/4/1855
Liverpool	44	*Anna*	25/4/1855
Plymouth	47	*War Cloud*	20/3/1855
Plymouth	48	*Conrad*	16/6/1855
Bristol	49	*Try*	14/5/1855
London	50	*Ballarat*	13/3/1855
London	51	*Cambodia*	13/7/1855
London	52	*Thomas Hamlin*	31/3/1855
Bristol	55	*Victory*	19/10/1861

On 5 May 1855, Beaufort records that 'Standard Marine Barometers' of the latest construction were dispatched by train to three navy ships as directed by FitzRoy, and instructions were given to suspend each one in a convenient place and register the readings from them instead of from the barometers with which they had formerly been supplied. It is evident that the well-organised FitzRoy had been galvanised into action, as he had been in those earlier days when refitting the *Beagle* for its important voyage. FitzRoy was a man who got things done – and in this case quickly.

Figure 3.1 shows a typical Adie marine barometer, most likely of the type issued in the early days of the Meteorological Office. Figure 3.2 is a close-up of the barometer in its transport case, showing the maker's label and Kew certificate. Scientific barometers were normally tested at Kew Observatory against a standard instrument, and, provided it was within a certain accuracy, a certificate was issued accordingly. Small copies of these certificates were glued into the lids of the respective carrying cases of the barometers and showed the small amount of compensation required at different readings to make the barometer correspond to the standard at Kew. This continued until the late 1990s when mercury barometers were no longer accepted for testing against the Kew standard.

Negretti and Zambra also supplied many instruments. This was a new company formed in 1850 as a partnership between Henry Negretti and

Figure 3.1 Marine barometer by Adie, Pall Mall, London.

Joseph Warren Zambra, who won medals for the instruments they displayed at the Great Exhibition at the Crystal Palace in 1851. The first record of a Negretti and Zambra instrument appears to be number 8, supplied on 17 April 1857, which was issued to HMS *Terror* on 20 July 1857. The Board of Trade also used instruments by Casella, although probably not until later. The issuing of new 'standard' barometers meant the recall of existing barometers, and about a hundred of these non-standard and less-reliable barometers were withdrawn from navy ships; of these, 58 were sold to opticians for use on an inferior class of ships, chiefly coasters. The rest were to be disposed of to the highest bidders, their value being applied in payment for new instruments of authorised construction.

It appears that both Adie and Negretti and Zambra were used extensively for the supply of barometers, and no doubt other instruments too, in the early years. FitzRoy is known to have visited Henry Negretti, and some correspondence between them survives. They had some discussion about the properties and capabilities of the curious 'storm glass' or 'camphor glass' (see chapter 5), among other topics.

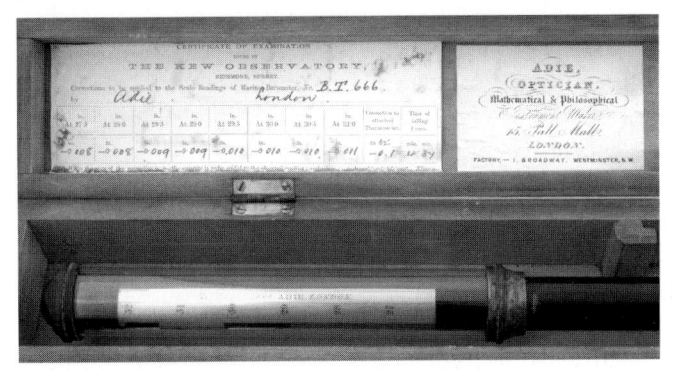

Figure 3.2 Detail of the head of the barometer illustrated in figure 3.1 in its transport case, showing maker's label and Kew certificate.

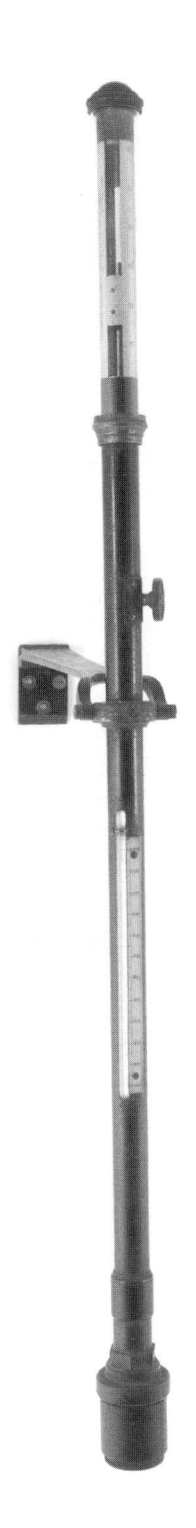

It is clear to see that, while the use of a barometer on board ship was a tremendous advantage in foretelling the weather at sea, carrying delicate glass tubes on board small timber ships was never going to be easy. Breakages were frequent, and there is constant reference to instruments being damaged and returned to the maker for repair. It is not surprising, therefore, that there was a need for a more robust type of instrument. The following is an extract from a letter from Captain Cockburn of HMS *Diadem* at Portsmouth, dated 23 October 1860:

> Sir, I have the honour to report to you that on several occasions the *Metal Barometer* supplied to the ship under my command has been broken, on two occasions by firing an accidental Gun, and on one occasion (which I witnessed myself) by a very slight blow or push.
>
> The very tenderness, or liability to break of this instrument, in my opinion renders this Barometer less useful than the wooden one – and for this reason I beg to be supplied with one of this description.

It appears that this letter was copied to Rear Admiral John E. Erskine of HMS *Edgar*, or sent to him and forwarded to the Board of Trade, though the records are not complete and we will never know the full details behind some of the correspondence in the archive; the Admiral was certainly aware of the facts. It is possible, however, that it was this letter that sparked FitzRoy into action, as can be seen by his reply to Captain Cockburn of 15 November 1860:

> Sir, I send such a barometer as you have asked for – but take on myself to send also a new kind of Marine Barometer, which, I hope, will answer for general public purposes better than any yet employed afloat.
>
> I direct this new instrument to Admiral Erskine – with a request that he will cause it to be tried in various ways, especially while Guns are fired near it on board the same ship?

In a separate letter of the same date to Admiral Erskine, FitzRoy wrote:

> Sir, I have the honour to send a Marine Barometer to you which is supposed to be sufficiently improved in its construction to be accurate – sensitive – free from 'pumping' – and yet able to bear the concussion of guns fired near it on board the same ship.
>
> May I request that you will cause it to be well tried, either on board the 'Diadem', or any other ship.

FitzRoy Marine or 'Gun' Barometers

Enclosed is a memorandum respecting its peculiarities.

We can presume that this barometer was the first of its type to be sent for trial, though the instrument, number 130, arrived damaged, along with its spare tube, during carriage from London. FitzRoy commented that it may have been some oversight in packing or neglect on the railway. Instrument number 107 was quickly sent to replace it and arrived intact. Interest must have been stirred, for Captain Cockburn and Captain Hancock also wished to receive one of these new marine barometers. FitzRoy obliged, stating that: 'It will be gratifying if it stands the concussion of gunfire shock.' Captain Wood of the *Barracouta* also requested one of the new instruments on 1 January 1861, so word was obviously getting round – but what of the tests?

We do not appear to have a record of all the correspondence concerning the barometer trials. On 30 January 1861, FitzRoy wrote a letter to Captain Superintendent Hewlett of HMS *Excellent* and the Royal Naval College at Portsmouth in which he acknowledges 'a valuable memorandum, respecting trials of Marine Barometers' and suggests further trial of several marine barometers at one time. These barometers 'may be suspended by their head-rings, or on their screws'. FitzRoy makes reference to previous trials: 'sixth of January 14th and the 1st of January 15th and to cease with the utmost trial that a Marine barometer may be required to stand onboard a Ship of War'. He suggests that only the Kew barometers with rubber packing should be tested and states that 'Mr Negretti' would be on hand with some of these new barometers and would, if required, show the readiest way in which tubes can be changed when necessary. The price of each new barometer with spare tube and carrying box was 4 guineas.

It can reasonably be imagined that several trials were needed, but these last ones on HMS *Excellent* appear to have been very successful. In Negretti and Zambra's *Treatise on Meteorological Instruments* (1864) they state:

> Some of the first barometers made by Messrs Negretti and Zambra on Admiral FitzRoy's principle were severely tried under the heaviest naval gun firing, on board HMS *Excellent*; and under all the circumstances, they withstood the concussion ... and the result was, according to the official report, 'that all these barometers, however suspended, would stand, without the slightest injury, the most severe concussion that they would ever be likely to experience in any sea-going man-of-war'.

It appears that Negretti and Zambra had successfully made a more robust barometer intended for general use, no doubt upon FitzRoy's request. Figure 3.3 illustrates two of their marine barometers from their catalogue of around 1880. On the left is a FitzRoy marine or gun barometer with ceramic scales, and on the right is the Kew marine barometer which was more generally issued by the Board of Trade.

As an interesting epilogue, we read in a letter of 18 September 1862 from Captain Hancock of HMS *Immortalite* that his barometer had broken during the firing of three blank rounds at night. It appears that it was dead calm and there was exceptional noise and concussion from the firing, so much so that it knocked pictures off the bulkheads and produced other damage. Captain Hancock suggests that exceptional atmospheric conditions may have been the cause. He also remarks that the spare tube was of great use. Being easily and readily fitted by himself, the barometer was working again in a short time and he was grateful that he had a functioning barometer by which to judge the weather. It seems that gunfire was not the only problem: FitzRoy was made aware of a barometer breaking by what was thought to be severe frost as it was positioned on the poop deck of HMS *Megora*.

Within Negretti and Zambra's *Treatise on Meteorological Instruments* is a clear description of this new improved marine barometer under the heading, 'The FitzRoy Marine Barometer':

Admiral FitzRoy deemed it desirable to construct a form of barometer as practically useful as possible for marine purposes. One that should be less delicate in structure than the Kew barometer, and not so finely graduated. One that could be set at a glance and read easily; that would be more likely to bear the common shocks unavoidable in a ship of war. Accordingly, the Admiral has devised a barometer, which he has thus described:–

'This marine barometer, for Her Majesty's service, is adapted to *general* purposes.

'It differs from barometers hitherto made in points of detail, rather than principle:– 1. The glass tube is packed with vulcanised india-rubber, which checks vibration from concussion; but does not hold it rigidly, or prevent expansion. 2. It does not oscillate (or pump), though extremely sensitive. 3. The scale is porcelain, *very legible*, and not liable to change. 4. There is no iron anywhere (*to rust*). 5. Every part can be unscrewed, examined, or cleaned by any careful person. 6. There is a *spare* tube, fixed in a cistern, filled with boiled mercury, and *marked* for adjustment in this, or *any similar* instrument.

FitzRoy Marine or 'Gun' Barometers

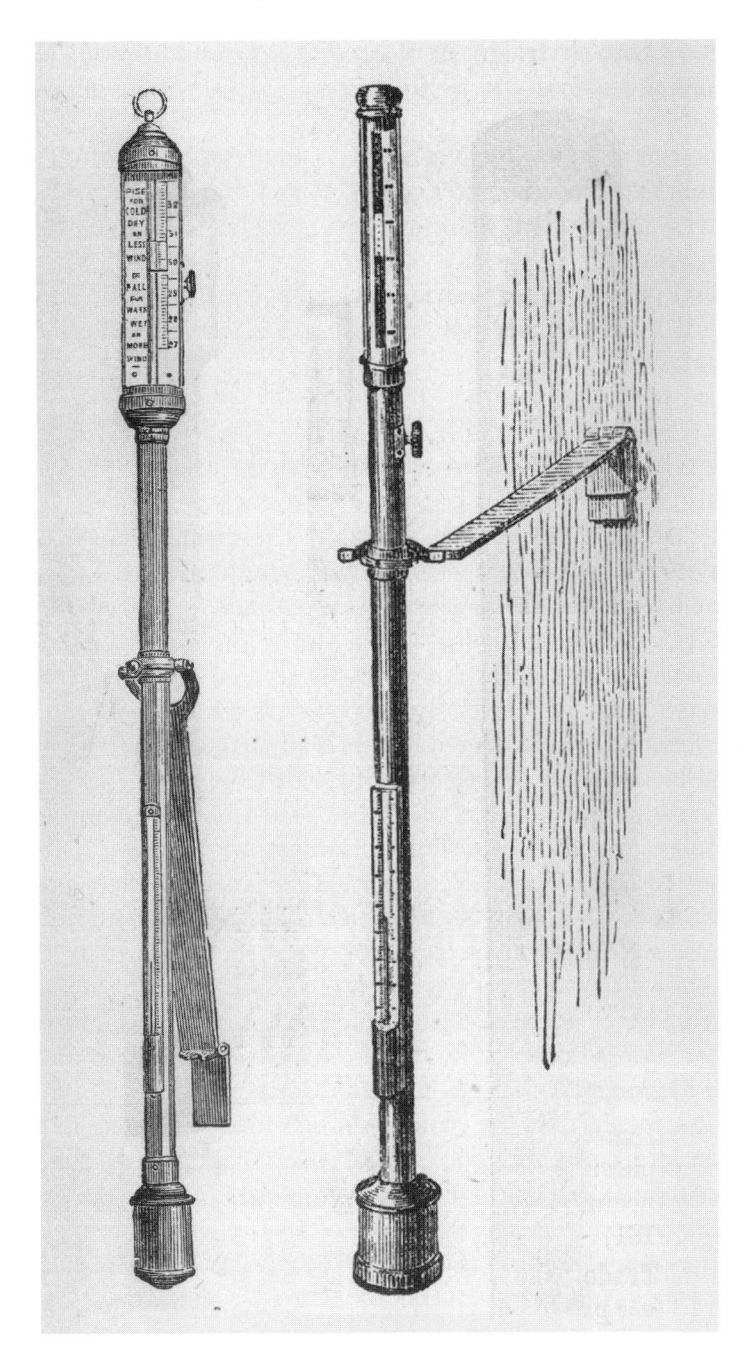

Figure 3.3 Brass gun marine barometer (left) and Kew pattern marine barometer (right) as illustrated in Negretti and Zambra's catalogue of around 1880.

'These barometers are graduated to hundredths, and they will be found accurate to *that* degree, namely the second decimal of an inch.

'They are packed with vulcanised caoutchouc, in order that (by this, and by a peculiar strength of glass tube) guns may be fired near these instruments without causing injury to them by ordinary concussion.

'It is hoped that all such instruments, for the public service at sea, will be quite similar, so that *any* spare tube will fit *any* barometer.'

Perhaps Negretti and Zambra hoped to corner the market, for if all barometers had the same-size tube they would presumably be the only makers, and the Royal Navy had need of many hundreds of barometers. But in more than 25 years of handling and restoring barometers, I have come across only two or three barometers that might fit the description of this type of barometer. There is one on display at Arlington Court, Devon, the ancestral home of the Chichester family (now administered by the National Trust), and it is certainly one of very few surviving. It is made of brass, but the brass has been coloured by a chemical method to avoid the constant cleaning that brass would normally need. This type of finish is something like blued steel but more green in colour.

The FitzRoy marine barometer chiefly varies from the more accurate and generally supplied Kew pattern marine in the following visible points. The top of the barometer is much wider to accommodate the ceramic scales bearing the words on the left and the pressure measurement on the right-hand scale. The vernier-adjusting knob is on the side of the barometer to the right of the scales and not below the scales. There is a hanging ring at the top so it can be used on land as well – something that most Kew pattern barometers did not continue to include, although in FitzRoy's time both models probably still had this feature. Generally, the scales are white ceramic with black lettering, although I have seen one in lacquered brass that had red and black letters. One remarkable such marine barometer we restored some years ago had a brass case in the pattern of a barley twist along its entire length (figure 3.4). It was an impressive piece for that method of design and manufacture alone; it also had silvered engraved scales with twin verniers by Adie of London, numbered 1182 (figure 3.5). It was most likely a specially made piece of slightly later date than we have been discussing here, perhaps dating from the 1890s.

FitzRoy gun marine barometers usually had words on the scales, which even then would have been uncommon for an accurate instrument as this

Figure 3.4 A curious 'barley twist', brass-cased marine barometer by Adie, no. 1182, c.1890. The bottom peg or spike is a type of locking system to stop the barometer swinging on board ship during travel and when not being read.

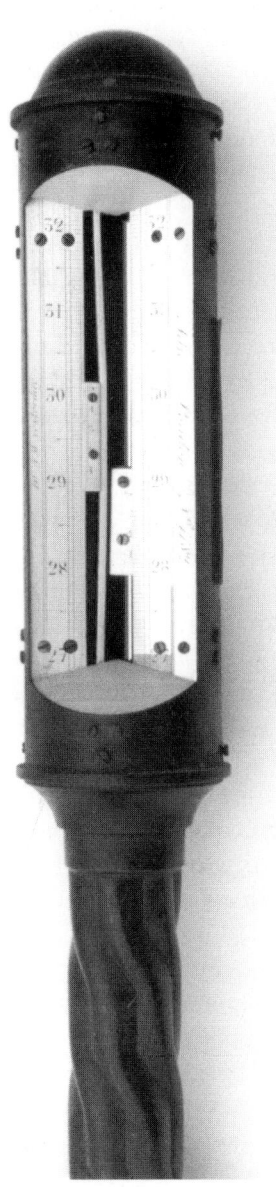

Figure 3.5 Top section of marine barometer illustrated in figure 3.4, showing the widened top scales with twin verniers and high-quality engraving.

might have led to some misunderstandings – but it was, as FitzRoy mentioned, for *'general* purposes'. Perhaps the design never really took off, or perhaps the instruments have just not survived. Many more Kew pattern barometers survive from a later period. It may be that, as metal-hulled ships took over from wooden ships and more stability was gained in the larger warships, and as gun design changed, the need for such instruments vanished, and they were replaced by the more popular and accurate Kew pattern barometer.

Figure 3.6 shows a Kew pattern marine barometer of the standard 1940 design, several of which still exist by different makers; this one is by F. Darton and Co. The barometer is shown in its transit box containing rubber and foam shock absorbers. There is foam at both ends of the box and the barometer is cradled in rubber at two points; the lid has two pieces of rubber screwed to it so that when the lid is closed the barometer is held securely in place. The box also has the original rope handles, the ends of which can be seen coming through the lid and screwed firmly in place; earlier boxes had a single rope handle. The handles are set off-centre to accommodate the heavy weight of the mercury reservoir and allow the barometer to be carried more or less level.

All Kew pattern barometers are made from one brass tube along the length of the barometer and the scale is divided very accurately at the top section so that, with temperature variations, the expansion and contraction of the scale can be calculated by tables. The Kew Observatory required that, to be accurate (that is, scientifically accurate), the scales had to be part of the body of the barometer and could not move or be moved. If, for instance, the scales were ceramic they would not operate in the same way and perhaps different tables would be needed to correct these more 'general purpose' barometers. The FitzRoy gun marine barometers, with their ceramic scales, would have been unlikely to pass Kew's strict test to obtain a certificate of accuracy: the scales were independent of the main body of the barometer and could move slightly or, if dismantled, may not be returned exactly to the original position; the scales were crudely divided as opposed to precision machine engraving with a vernier that could be read to an accuracy of a hundredth of an inch.

In the Board of Trade register of barometers issued there is mention of the issue of 'New Pattern' barometers, but these only occur for three pages of the register from early 1861; after that, no more reference appears. Either they were all 'new pattern' from then on (and therefore not specially designated) or else they reverted to the old style, which is more probable. There is still mention of many barometers by Adie and a few by J. J. Hicks (another eminent instrument-maker who began making in 1861), which

FitzRoy Marine or 'Gun' Barometers

suggests that FitzRoy's style was no longer suitable. The end of his influence in the choice of instruments consequent upon his death in 1865 may have been a major factor in the apparent discontinuation of these barometers in general naval service, although Negretti and Zambra were still advertising them in their 1913 catalogue, as can be seen in figure 3.7, which shows the marine barometer commonly in use on ships, priced at £4 10s (left), as well as 'Admiral FitzRoy's' gun marine barometer priced at £5 10s (right). Perhaps the real reason why FitzRoy's gun marine was less popular was this price differential. Also illustrated in figure 3.7 is a Kew barometer (middle) mounted on a wooden board for use on land.

In a Casella catalogue of around 1920 (figure 3.8), the Adie-style marine is advertised. By the 1930s, Negretti and Zambra had dropped the FitzRoy gun marine and were only advertising the conventional ship's marine as can be seen in figure 3.9. By this time, the cast-iron reservoir has changed to a simple, straight-sided, turned metal reservoir, probably stainless steel, although it is difficult to tell from the illustration.

Figure 3.6 Kew pattern marine barometer of the 1940 type in transport case by F. Darton and Co.

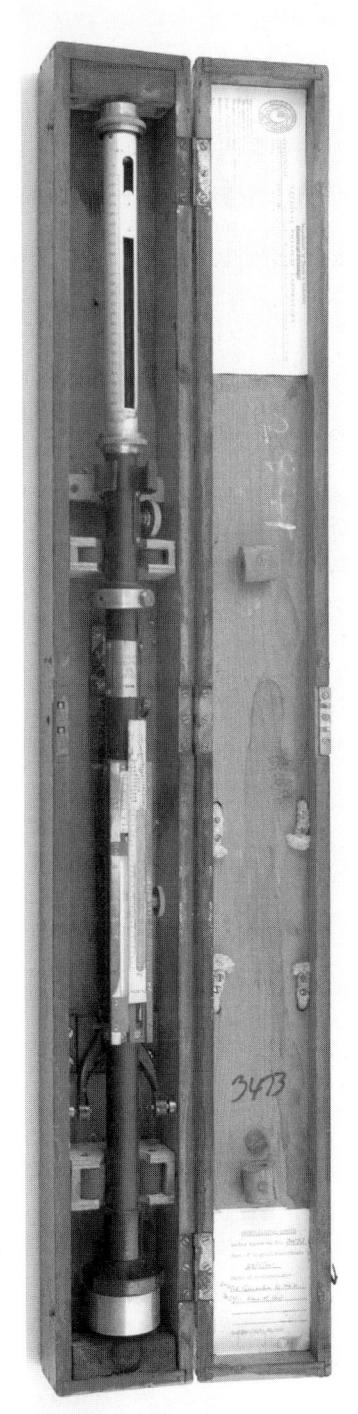

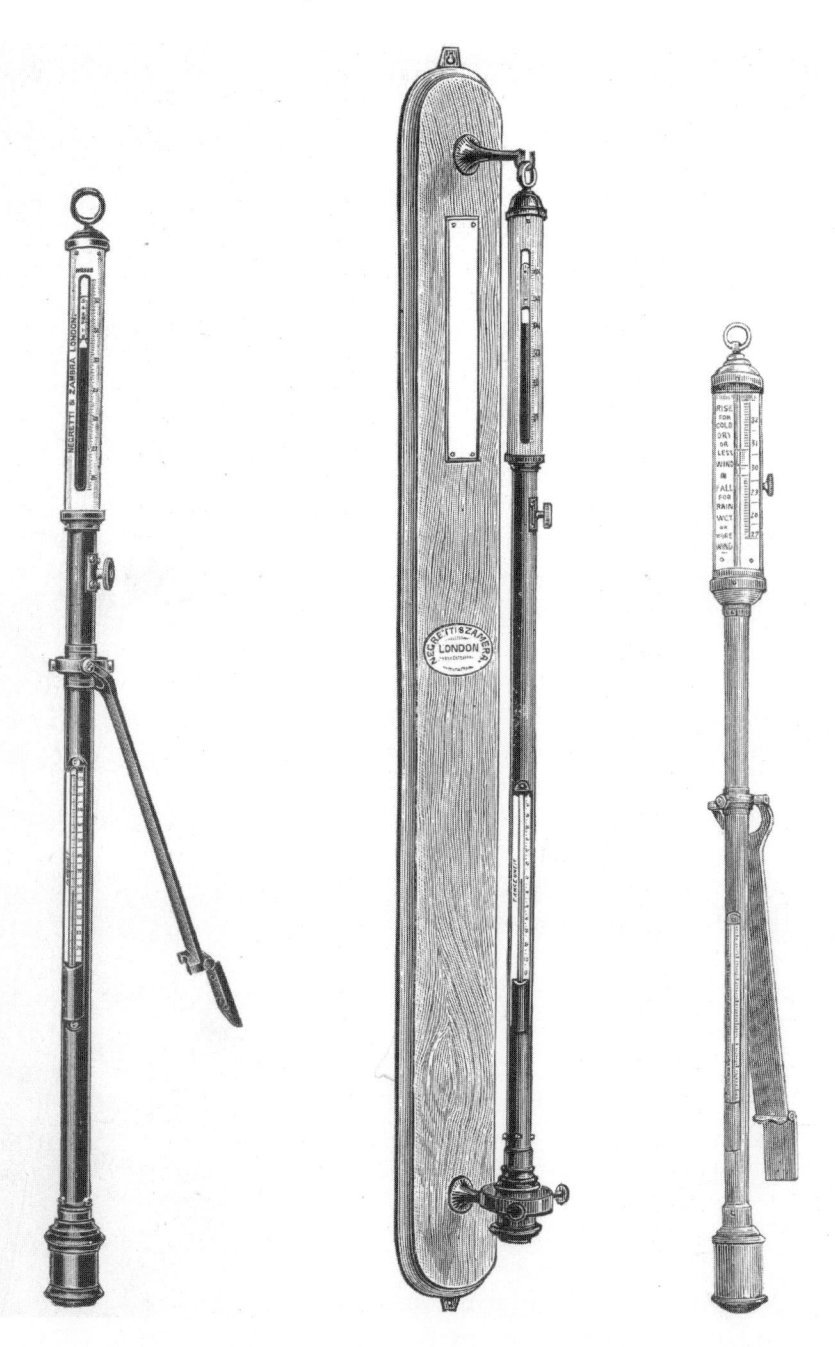

Figure 3.7 Marine barometer (left), station Kew barometer for land use on wooden back board (middle), and 'Admiral FitzRoy' gun marine (right) as illustrated in Negretti and Zambra's catalogue of 1913.

Figure 3.8 Adie-style marine barometer illustrated in Casella catalogue number 498, c.1920.

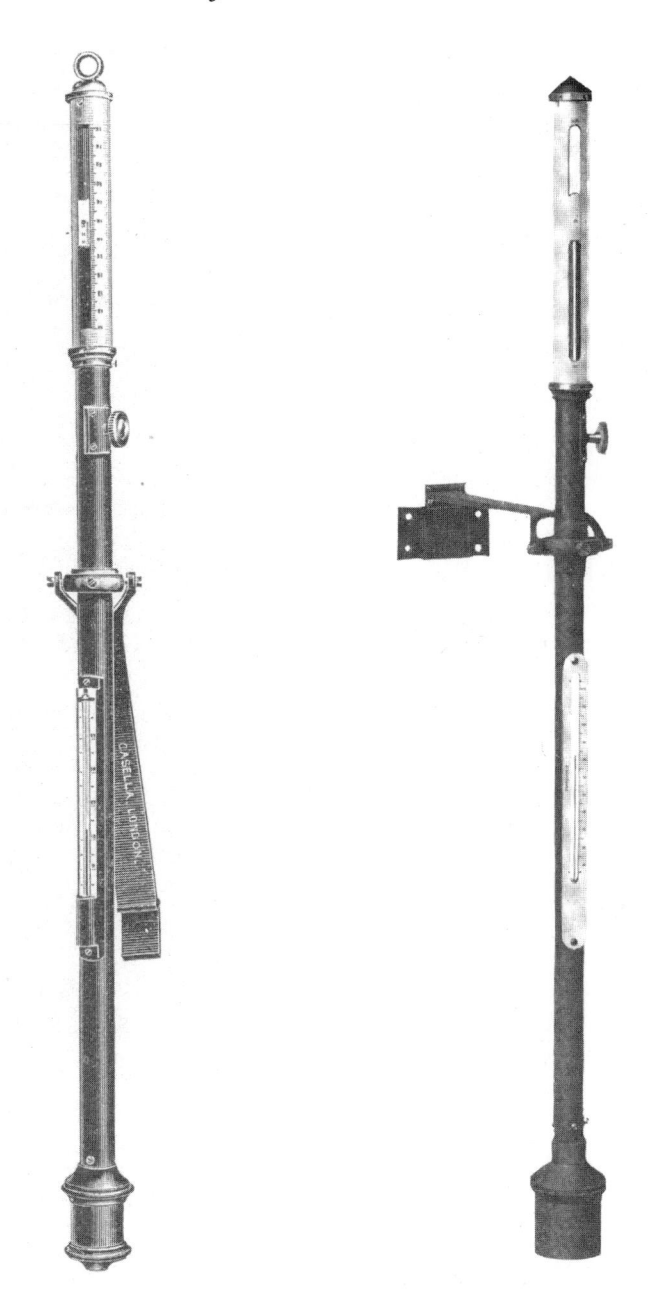

Figure 3.9 Marine barometer illustrated in Negretti and Zambra's catalogue (E5) from the early 1930s.

4 FitzRoy Storm Barometers

Early in his new position at the Meteorological Office, as we saw in chapter 3, FitzRoy arranged for barometers to be issued to ships and some stations to obtain readings and thereby statistics to begin to understand the weather. In 1857, he also considered having barometers placed around the British coast. These barometers began life as 'fisheries' or 'coastal' barometers, and a few early examples have a porcelain plate above the register scales marked 'sea coast barometer', but they later became known as 'FitzRoy storm barometers'. For consistency, they will generally be referred to here as storm barometers, except where original documents and references state otherwise.

With the idea of supplying barometers to coastal regions, on 9 December 1857 FitzRoy wrote to Negretti and Zambra, a firm of meteorological instrument-makers already known to him for the manufacture of Kew pattern marine barometers, with the following proposal:

> It is intended to place ordinary land barometers, as *weather glasses* solely, at *some* of the more exposed Fishing stations, and coasting harbours, in Great Britain and Ireland.
>
> I wish to see and confer with you on this subject and will thank you to fix an hour – between one o'clock and three – as soon as one of you can make it convenient to call here? It would tend to simplify and forward arrangements if you would bring a few specimens of such instruments as seem to you likely to meet the following requirements.
>
> 1 The barometer may be in a wooden case – with a leather cistern.
> 2 It should have a pipette in the tube: but no other contraction.
> 3 The range of *ivory* scale should be from 27 to 31 inches adequately *verified* by testing.
> 4 It should have a transport screw and read to *hundredths*.
> 5 The attached thermometer should be good and legible.
> 6 It should be so packed in a stout roomy case, with springs or otherwise, that it will bear transport to the Isles of Scotland – or elsewhere – by sea and land – provided reasonable care be taken.

And after providing as above mentioned, under the six heads – the total charge, for *all* but carriage must be as low as possible consistently with the Maker's fair profit – and prompt payment of HM Government.

It can easily be seen that FitzRoy, with his practical eye and knowledge of barometers, had formed a clear idea of the type of instrument he required. Just how many and what style of sample barometers Henry Negretti or Joseph Warren Zambra brought to FitzRoy's office in response to this request, we will never know. No doubt, as with the marine barometers, a few variations were tried, but we have many surviving examples of barometers to show the design that was finally agreed upon.

Although no mention was made in the letter to Negretti and Zambra about the words to be put on the scales of FitzRoy's storm barometers, in his sixpenny *Barometer Manual*, issued by the Board of Trade, FitzRoy prints an 'Explanatory of Weather Glasses in North Latitude' (figure 4.1), helpfully summarised as four simple columns at the foot of the page, which he notes 'may be useful' on barometer scales. Either FitzRoy was already requesting this to be produced on his storm barometers or he was considering it. I suspect that he had already designed it, and the *Barometer Manual* coincides with the barometers being issued to the fishery stations around the British Isles.

In February 1857, Joseph Glover and John Bold of Liverpool filed a patent, number 501, relating to 'Improvements consisting of extended uses of photography as applied to dials, tablets and pictures', which was essentially a method of preparing clear glass, opal glass and other substances to accept photographically produced designs and then finally coating the picture or impression with transparent varnish. Perhaps Negretti and Zambra could see the advantage of such a method and refined it further, as in October 1857 they filed patent number 2641 for 'Improvements in producing graduated scales and other signs, letters, numerals, characters, and pictorial representations upon porcelain and other ceramic and enamelled materials, which improvements are applicable to the graduated scales of meteorological and other philosophical instruments'. Negretti and Zambra's patented method is similar to the earlier patent, but instead of producing the final design photographically, they used the process to chemically etch the design onto the material.

During early 1858, FitzRoy was writing about porcelain scales to a Mr Welsh, who had found porcelain advantageous over metal scales. Instead of ivory for the barometer scales, porcelain could be used, which is clearly legible and extremely resistant to corrosion and wear. No doubt porcelain

EXPLANATORY OF
WEATHER GLASSES
IN NORTH LATITUDE.

IN OTHER LATITUDES SUBSTITUTE THE WORD SOUTH, OR SOUTHERLY OR SOUTHWARD, FOR NORTH, &c. THROUGHOUT THESE PAGES.

THE BAROMETER RISES for Northerly wind,	THE BAROMETER FALLS for Southerly wind,
(including from North-west, by the *North*, to the Eastward,)	(including from South-east, by the *South*, to the Westward,)
for dry, or less wet weather,—for less wind,—or for more than one of these changes:—	for wet weather,—for stronger wind,—or for more than one of these changes:—
EXCEPT on a few occasions when rain (or snow) comes from the Northward with *strong* wind.	EXCEPT on a few occasions when *moderate* wind with rain (or snow) comes from the Northward.
For change of wind towards *any* of the above directions:—	For change of wind towards the *upper* directions *only*:—
A THERMOMETER FALLS.	A THERMOMETER RISES.

Moisture, or dampness, in the air (shown by a hygrometer), increases BEFORE or with rain, fog, or dew

On barometer scales the following contractions may be useful in *North* latitude:— And the following Summary may be useful *generally*:—

RISE FOR N. ELY. NW.–N.–E. DRY OR LESS WIND. EXCEPT WET FROM N. ED.	FALL FOR S. WLY. SE.–S.–W. WET OR MORE WIND. EXCEPT WET FROM N. ED.	RISE FOR COLD DRY OR LESS WIND. EXCEPT WET FROM COLD SIDE.	FALL FOR WARM WET OR MORE WIND. EXCEPT WET FROM COLD SIDE.

Figure 4.1   FitzRoy's 'Explanatory of Weather Glasses in North Latitude' from his *Barometer Manual*, p. 3.

would have been cheaper to produce than ivory, which in those days was mostly hand engraved. These new scales, once fired and glazed, were chemically etched, and this etching then filled with black paint or similar. On some early models, I have noticed that the black filling has gradually come out, but can often be replaced satisfactorily.

Many storm barometers have a plate that states: 'This barometer reads correct with the Greenwich Standard.' In early barometers, this plate is made of glass, blackened on the reverse and with white lettering, and this is clearly produced by Negretti and Zambra's patent process. As with the ceramic scales produced with this method, the white filling becomes loose with age and is often missing or in the process of falling out. These types of plate are signed by James Glaisher FRS. Glaisher was superintendent of the Magnetic and Meteorological Department of the Royal Observatory from 1838 until he retired in 1874. Later models – certainly therefore after 1874 – do not have a signature, and even later models do not have any type of plate at all. The later plates changed from a blackened glass plate to a white glass plate with black lettering, using transfer-fired scales which are totally stable and permanently fixed. (White glass was a material increasingly used in barometers in the late Victorian period for stick barometer scales.) It is difficult to discover exactly when the use of a plate was discontinued. I would estimate certainly by 1900, although the domestic models illustrated in Negretti and Zambra's catalogue of 1880 do not have this plate at all.

On more expensive models sold by Negretti and Zambra, you see examples of more elaborate scales, sometimes with red-coloured letters and ivory or even silvered brass scales. The choice for the customer was open to variation. Only for the Board of Trade was there no need to vary a satisfactory design: each order would have been for the same instruments as previously supplied, unless some modification was required for accuracy, transportation purposes or other operational improvement.

Once decided upon, it seems that there was very little change in design from FitzRoy's original requirements, perhaps with the exception of the top of the barometer, which was sometimes flat or angular in some early models (c.1860), but changed to the more commonly seen half-round style. I suspect that both types were made as required and so no clear date can be inferred by one type or the other. Figure 4.2 is an illustration from Negretti and Zambra's catalogue of instruments of around 1880. The barometer illustrated is probably from an existing engraved printing plate and is titled 'Sea Coast Barometer by Negretti & Zambra, London'. This is certainly from the first design of barometers that FitzRoy requested, being 'Sea Coast Barometers'. As the design of the case did not change much, they have used an earlier illustration. (Bearing in mind that each illustration was hand engraved and therefore quite costly, many engraved printing plates were kept and used in later catalogues.)

Figure 4.3 shows the actual plate from a similar barometer, with the added words 'as used by the Royal National Life Boat Institution'. I think Negretti and Zambra sold barometers with this wording as a form of marketing, a way of promoting the barometer as being a good instrument. Very few barometers actually marked 'Sea Coast Barometer' survive, which is another indication of their earlier date. Invariably, the barometers that survive tend to be marked 'Admiral FitzRoy's Storm Barometer' and they date from after FitzRoy's death in 1865.

The text in the Negretti and Zambra catalogue lists 'The Fitz-Roy Storm Barometer, or Fisherman's and Life Boat Station Barometer, as made by Negretti and Zambra especially for the Board of Trade and Royal Life Boat Institution, to be fixed at all the principal Seaports, Fishing and Life Boat Stations. Price, £5 5s.' The same design with two verniers was £6 10s, and with a carved, ornamental case £8 8s. Figure 4.4 shows a typical FitzRoy storm barometer in position at a small coastal village – probably of a much later date as they were still being issued well into the twentieth century, often by public subscription but also by the Meteorological Office and some wealthy individuals.

In June 1858, FitzRoy requested 'permission to print 1,000 copies of his Barometer Manual, Weather Guide and Card for gratuitous distribution in poor places, and for sale elsewhere (at sixpence each)'. The request was sanctioned, but with a warning from the President of the Board of Trade, the MP Joseph Warner Henley, that 'it is not to be taken as a Precedent for incurring further expense

Figure 4.2 'Sea Coast Barometer' as illustrated in Negretti and Zambra's catalogue of around 1880.

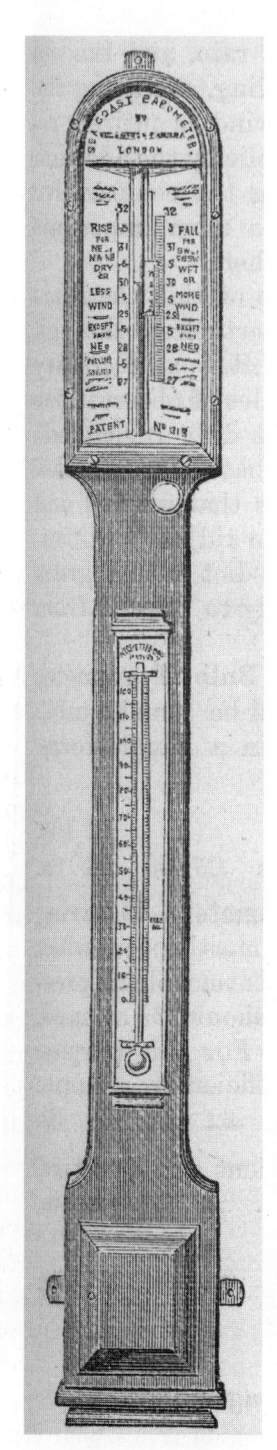

Figure 4.3 Plate from a 'Sea Coast Barometer' by Negretti and Zambra.

for setting up, printing, or publishing any further paper.' The cost of printing was £24 5s, so there was no profit at 6d each and many were given away.

In one letter from FitzRoy, he suggests that 'practical fishermen' would benefit from the *Barometer Manual* being printed in large type. (In fact, he amended the word 'practical' from 'simple', which was originally written in a draft letter, obviously being sensitive to calling fishermen 'simple' but aware of their limited education at that period.) The *Manual* is in very large, easy-to-read print, and was often sent along with a barometer when being issued. FitzRoy must have empathised with the practical men who were being taught to use the barometer; he had no doubt learnt from his years on the *Beagle* how to teach his crew in the most effective manner.

The *Barometer Manual* is a short guide of 32 pages. The first two pages of text give brief directions for the use of the barometer (the second page is reproduced in figure 4.1 above). There then follow 92 numbered paragraphs under the heading 'How to Foretell the Weather', which give useful observations by which 'any one not accustomed to use a barometer may do so without difficulty'. As well as guidance on how to 'read' a barometer, taking into account temperature, wind direction and force, FitzRoy also includes more homespun 'weather wisdom' in the belief that 'every prudent person will combine observation of the elements with such indications as he may obtain from instruments'.

It is clear that early storm barometers were issued and logged according to the manufacturer's instrument number as was common with marine barometers, but at some stage this must have changed. There is at least one surviving example which has Negretti and Zambra's instrument

Figure 4.4 Storm barometer in position in a coastal village.

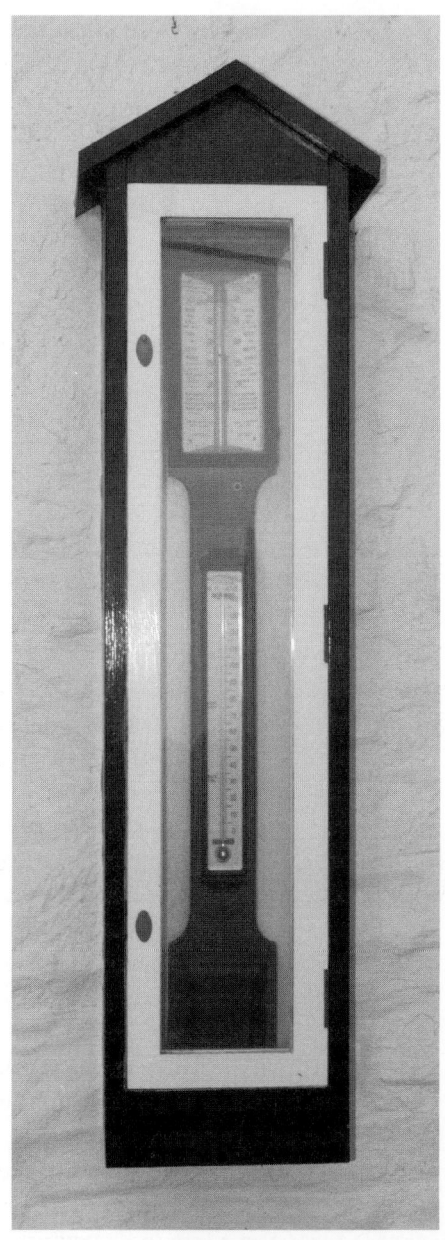

Figure 4.5 Storm barometer by Negretti and Zambra in original coastal position.

number 2367 on the right-hand scale, while on the left scale, towards the top, is 'MO 218', which is the Meteorological Office's reference for the instrument, and on the thermometer, which is mounted on the trunk of the barometer, there is 'MO 4735'. (The barometer is shown in figure 4.5, though these numbers are not visible in the picture.) Clearly, the Meteorological Office considered (rightly) that for their records they had issued two instruments, one barometer and one thermometer; the numbers indicate that they had issued many more thermometers than barometers by the time the barometer in question was issued. I have not seen another like it and suggest that it is indeed a relatively late method of marking which may be very late Victorian or, perhaps more likely, early twentieth century.

Not all cases were of the rounded pediment design. There are also a number of flat top or 'caddy' pediment ones, such as that shown in figure 4.5. This instrument, in a small fishing village in the south west, is housed in a simple glazed box, and many were displayed like this or on occasions the box was built into the wall. Several other variations creep into the design: sometimes the top name plate is produced without the common words 'FitzRoy Storm Barometer' and 'Negretti & Zambra Instrument Makers to Her Majesty, London.' No doubt, over many years in production, the design changed in minor ways, and especially for those that were sold to institutions and private customers. Then there are other makers who copied the design to a greater or lesser degree. The variations that can be found will be subtle but to the trained eye quite noticeable.

The cases of all storm barometers for public display were made of oak as it is a strong, durable wood that weathers well. Examples that I have handled are clearly weather beaten, the grain having been worn away over many years of harsh conditions near the sea. This shows that the barometers were often in very exposed situations, though weathering can be a good point in a barometer as it testifies to its many years of coastal use. These barometers, however, even the weather-beaten ones, were not hung up unprotected. They were housed in stout cases with lockable glass fronts. One still in existence is also behind a closed solid wood door (see figure 4.4), so perhaps only local people would know where to consult it. Most often, the barometers were clearly visible, and twice daily the barometer would be read, the indicators adjusted, the issued charts marked and placed on public display, and the information telegraphed (in the early days) back to FitzRoy's head office.

According to the advertisement in Negretti and Zambra's catalogue, the FitzRoy storm barometer:

consists of a tube with very large bore, and an accurate Thermometer, mounted in a solid oak frame, firmly screwed together [this relates to the brass screws that hold the wood together so that if the glue gets damp it does not fall apart], with scales and figures, &c., permanently engraved on *Porcelain*, by Negretti & Zambra's Patent process, the Vernier reading to 100-ths of an inch. It is strongly recommended as a good, sound working instrument, admirably adapted for use in Public Institutions.

In 1859, FitzRoy was elected as a member of the Committee of Management of the Royal National Lifeboat Institution (he paid £1 subscription annually). This organisation, with its primary aim of saving lives from shipwreck, had begun in 1824 after an appeal to the nation by Sir William Hillary. The first lifeboat service was called the National Institution for the Preservation of Life from Shipwreck, but in 1854, the same year as the Meteorological Office was formed, it was re-named the Royal National Lifeboat Institution (RNLI).

FitzRoy's distribution of public barometers around the coast amongst the poorest and most exposed fishing communities would have been well known within the RNLI. In June 1860, the Committee of the RNLI requested FitzRoy's cooperation in establishing 'Barometers with instructions for their use at Life-boat stations', a request that he wholeheartedly supported. He replied on 19 June 1860 and sent a storm barometer, packed in its special box, for the inspection of the Committee. FitzRoy's letter ends: 'As a member of the committee of the Institution, I shall have great satisfaction in aiding its objects – the utility of which cannot be disproved or over-rated.'

Quite where the drive came from for this initiative is not easy to ascertain: was FitzRoy's recent election to the Committee part of the RNLI's strategy to save lives? Or was it engineered by FitzRoy to further the spread of barometers around the coast? Perhaps FitzRoy suggested the positioning of barometers at RNLI stations because of a lack of funding by the government. In the October 1858 issue of the RNLI's journal *The Life-boat*, we read that FitzRoy had written to the RNLI extolling the virtues of the barometer, and enclosing some copies of his *Barometer Manual*; he explained about the provision of barometers to some of the poorer fishing villages around the coast with the assistance of the government. In the *Illustrated Times* of 3 November 1860, it was reported that the RNLI was to provide all of its lifeboat stations with a barometer as per the Board of Trade's design. In 1861, there were 110 lifeboat houses.

By the end of 1861, at least 35 storm barometers had been allocated to lifeboat stations, or other useful public places, by the RNLI to aid in the preservation of life around the coasts. By 1878, almost 200 barometers had been provided by the RNLI. The records show a very marked reduction in the provision of mercury barometers after this time, until from 1900 only about six were recorded as being issued, and some of these were probably replacements for damaged ones already in existence, the last one being listed in 1930.

All storm barometers issued by the RNLI up to number 118 in 1876 were sent to Greenwich for verification; thereafter, they were supplied direct from the workshops of Negretti and Zambra. These barometers were not only provided for lifeboat men but sometimes for fishermen and for public use. Sometimes the instrument was housed inside the window of a lifeboat station; one that I have seen could be turned simply by pulling a cord (about the size of a sash cord) to allow the barometer to be read outside the lifeboat house or used by the lifeboat men inside when needed. More normally, it was positioned in a well-frequented public place so as to be visible from outside either in a window or a specially made glazed box.

Perhaps the person most responsible for the supply of barometers by the RNLI to their lifeboat stations was Algernon Percy, 4th Duke of Northumberland. According to an article by James Glaisher in *The Lifeboat* of 1 January 1861, the duke proposed as early as 1859 to the president of the British Meteorological Society the establishment of meteorological observations at several fishing villages on the coast of Northumberland. The primary object of the duke was the saving of life and property. He paid for 14 barometers to be situated in public places, including lifeboat houses, along the Northumberland coast.

These barometers appear to be somewhat different in design from the FitzRoy storm barometer: they show none of FitzRoy's scale words, have an angular pediment, a thermometer alongside the barometer scale and a plate which reads 'Northumberland Coast Station Barometer No. 1 [to 14]. Established by His Grace The Duke of Northumberland and the British Meteorological Society. Barometer reads correctly with Greenwich Standard Sept. 1860. Jas. Glaisher. F.R.S.' (figure 4.6). I suspect that several others of a similar type were issued in later years. According to Negretti and Zambra's 1880 catalogue, other public-spirited people also supplied them, and later some barometers were paid for by public subscription and owned by the parish, presumably where the Board of Trade would not issue them free of charge.

RNLI storm barometers seldom vary except for the mounting arrangement, which is invariably with heavy cast brackets, and the name

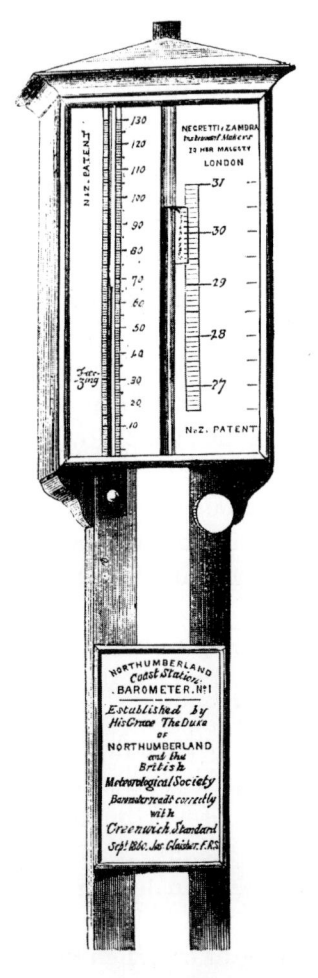

Figure 4.6 'Northumberland Coast Station Barometer' No. 1, by Negretti and Zambra (from *The Life-boat*, 1 January 1861, p. 368, by courtesy of the RNLI).

plate, which is numbered as in figure 4.7 (here no. 73, c.1865). This number relates to the RNLI's own numbering system which is different from the instrument-maker's number (which appears on the lower part of the right-hand barometer scale). The glazing is crazed as often happens with old porcelain scales but seldom as pronounced as seen in this example. Figure 4.8 (the plate for no. 187) has similar wording but shows a different style and size of lettering, which is found on most of the surviving barometers from the last issue by the RNLI of around 1878. Private or publicly subscribed barometers often had name plates of a similar form, but not numbered and perhaps stating 'as used by the RNLI' (as figure 4.3), though the later ones would be without the description 'Sea Coast Barometer'. Many RNLI barometers survive in lifeboat buildings and offices today, some being removed to a convenient and lockable harbour master's office for safety.

It seems that FitzRoy considered that the severe weather conditions around the Scottish coast were of greater threat to the fishing industry than elsewhere, and so the first ten 'Fisheries Barometers' were dispatched in June 1858, nine (including one spare) to Scottish fisheries and one to St Ives in Cornwall. Interestingly, it was George James Stebbing, the instrument-maker who had travelled with FitzRoy on the *Beagle*, who was entrusted to deliver and set up the Scottish barometers, first eight, then another eight. FitzRoy was charged by the accountant H. R. Williams with incurring higher than necessary costs in transporting these barometers. Fitzroy justified these and other costs on 6 August 1858 in a 16-page letter. He explained that these barometers were large –

having heavy cisterns full of quicksilver – and are not portable like

Figure 4.7 Name plate from RNLI barometer numbered 73, c.1865.

Figure 4.8 Name plate from RNLI barometer numbered 187, c.1878.

marine travelling barometers. (You can see one in my office at anytime.) If they were sent by common conveyance, even by water, without a special agent in charge, not one would arrive at its destination unbroken. Not only is a person required to guard them, but to place, and explain how to manage those instruments ... Twenty three barometers have been thus safely deposited *without a casualty* – from Lerwick to the Land's End at a total cost of five pounds each barometer including *all expenses* of conveyance. I should have said before hand – such success was not to be anticipated.

FitzRoy continues by explaining that he intended to draw up a paper for parliament next year which he hoped would have the effect of furnishing about one hundred places in the British Isles with these 'life-preserving instruments', besides the 30 already contemplated. It appears that the cost of sending the first few barometers to Land's End, St Ives and Leith in Scotland (perhaps others en route) by George James Stebbing was £13.

Word swiftly got round about the possibility of barometers being distributed, and soon private requests for barometers were received from a variety of coastal towns and villages. By April 1863, 89 'Fisheries Barometers' had been issued in England, Scotland, Ireland, Wales and the Isle of Man: 64 were 'loaned' by the Board of Trade and the rest were supplied as private gifts. Part of the design specification as originally set out by FitzRoy was for each barometer to be supplied in a stout box fitted with springs to make it safe to transport. Whilst none of these boxes has come to my attention, it is clear that there was generally a high success rate in delivering the barometers safely. In a letter to Sir James Emmerson Tennent, dated 11 August 1864, FitzRoy writes: 'Your barometer is under repair, and will be sent to you (when duly examined) in about a fortnight ... Out of about 80 such instruments sent to various stations between the Shetland Isles and Penzance – only one, besides yours, has been injured in travelling ...'.

As knowledge of weather forecasting grew and the value of a barometer to a coastal community became established, many fisheries (and they were numerous around Britain's coasts at this period) desired one; there was perhaps also rivalry between fishing communities to obtain a barometer, or jealousy when a neighbouring settlement received one. But it was not all plain sailing for FitzRoy in his attempt to supply barometers to these communities. Much work had to be done by FitzRoy to achieve his aims. He had to obtain permission to distribute the barometers, finance was always a problem, and he had a host of correspondence and work to undertake as well as overcoming some people's resistance. For instance, there was an old fisherman's superstition that even to consult a barometer was 'tempting Providence'. There were sometimes also suspicions between fishermen and coastguards which had to be overcome in order for barometers to be placed and used successfully. Coastguards were often suspected by fishermen of being informers about smuggling, and fishermen with dirty boots were not always welcomed by coastguards on their premises. FitzRoy's path was far from easy, but he was determined and industrious.

The intention behind the distribution of these barometers was twofold: first, weather information could be telegraphed back to FitzRoy's office for the purposes of statistical research and storm warning; and, secondly, local inhabitants such as fishermen could have access to a barometer, which they would not have had a chance of owning or using otherwise as only the wealthy could afford such items. This second intention, however, was not always realised in practice. There survive in FitzRoy's papers a number of letters from August 1861 concerning a barometer sent to a Mr John Carahur for the use of shipping frequenting the port of Dundalk in Ireland. The Royal Prussian and Danish Consulate in Dundalk informed FitzRoy that the barometer had been installed in a room used by Mr Carahur in his 'Public House' and that none could see it without entering those premises. It was suggested that it should be removed to the customs house where it could be placed in a window for all passing to see. FitzRoy agreed with the suggestion, obviously disturbed that it was in a 'Public House' (and no doubt aware what the pubs were like in British ports of the day) – which was far from his intention of 'public display'. This was, however, probably an exception, and gradually these barometers were distributed across Britain and occasionally further afield. The Dundalk barometer was destroyed during the Second World War.

Figure 4.9 is an example of a storm barometer in a decoratively carved case with flashed opal glass scales. Figure 4.10 gives a close-up view of the scales, which have red and black lettering to enhance its decorative appeal. There were many variations for those who could afford a more attractive barometer either for their home or for an impressive public building. They are not the ones that would have been used for FitzRoy's systems of weather forecasting and storm warning, but were sold by Messrs Negretti and Zambra to anyone who desired to have a similar barometer. FitzRoy and his system of weather forecasting remained famous long after his death, and Negretti and Zambra certainly appear to have used his name to their advantage both during his lifetime and after his death.

After FitzRoy's untimely death in 1865, Thomas Henry Babington took over the reins of the Meteorological Office. On 13 June 1866, just over a year after FitzRoy's death, Babington wrote the following letter to Negretti and Zambra, clearly annoyed at their 'marketing' methods:

> Gentlemen, My attention having been called to an advertisement on the Wrapper of a small work just published by Messrs Routledge entitled 'A Manual of Weathercasts and Storm Prognostics' – I feel it my duty; as, for the present, Chief of this Department, to observe that there are several statements or implications in that advertisement which appear to me to be unwarranted and which I am quite sure, in justice to others, Admiral FitzRoy would not have allowed to pass without remonstrance.
>
> Your advertisement *implies* that *all* the Barometers used by the

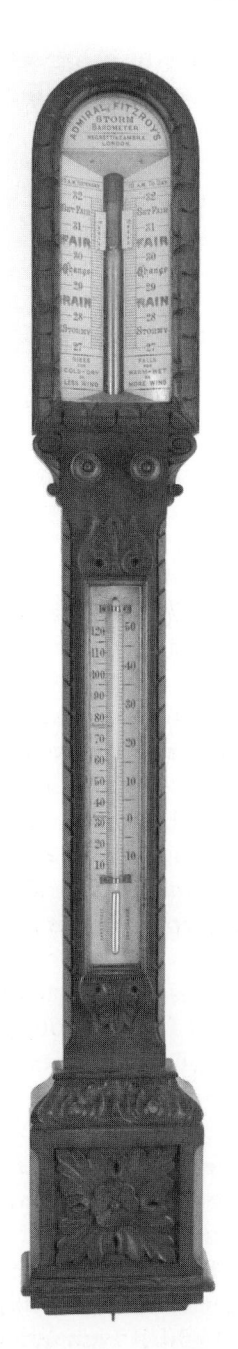

Figure 4.9 Carved oak FitzRoy storm barometer by Negretti and Zambra, c.1890.

Figure 4.10 Barometer scale of storm barometer shown in figure 4.9.

late Admiral FitzRoy for foretelling weather were made by yourselves and that the absurd title of "*Storm*" Barometer (inferring the possession of some property peculiar to an instrument so called, which is not possessed by other *well constructed* barometers) originated with Admiral FitzRoy. Both these implications are, as you must be aware, contrary to the fact.

What is meant by the statement that 'the only authentic Instruments' (designed by Admiral FitzRoy) are to be obtained etc. and that 'all others *are spurious and not recognised by the Board of Trade* etc. ... I am at a loss to understand.

We have no cognizance here of any barometer designed by Admiral FitzRoy.

Important *improvements* which have been adopted by yourselves in common with other first class makers – were doubtless suggested by him.

This does not appear to have halted Negretti and Zambra as the following extract from their 1880 catalogue shows: 'Messrs Negretti and Zambra would specially caution the Public against purchasing cheap and worthless imitations of Admiral Fitz-Roy's Barometers as leading to disappointment. Full details both as to the construction and use of the true Fitz-Roy instrument will be found in Negretti and Zambra's Barometer Manual, compiled by Admiral Fitz-Roy for the Board of Trade; post free, 6d.'

Misunderstanding over these FitzRoy barometers, originally called fishery or coastal barometers, and designed to be robust and simple to use, probably arises from Negretti and Zambra's marketing strategy. Negretti and Zambra were clearly the 'culprits' in marketing the barometer after FitzRoy's death and re-naming it – perhaps reflecting popular use at that time – the FitzRoy Storm Barometer. These barometers specifically vary from the more commonly found domestic 'FitzRoy barometer' to be discussed in chapter 5.

Figure 4.11 is a very typical storm barometer by Negretti and Zambra, with single vernier (damaged), rounded top, and long thermometer. The thermometer has much finer text than the scales, as can be seen in figure 4.12, and is no doubt of the type that was photographically etched; it was probably a leftover from existing stock as the address on the thermometer is 11 Hatton Garden which Negretti and Zambra left in 1859, whereas the scales (figure 4.13) have their address as 122 Regent Street, which became their head office in 1862. The barometer can reasonably be dated within the 1860s; it is numbered 378.

Figure 4.14 is a very similar barometer, numbered 1449, dating from

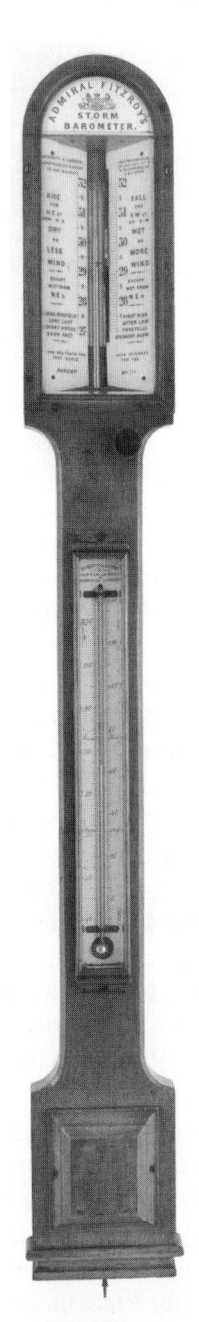

Figure 4.11 Storm barometer, number 378, c.1865, by Negretti and Zambra.

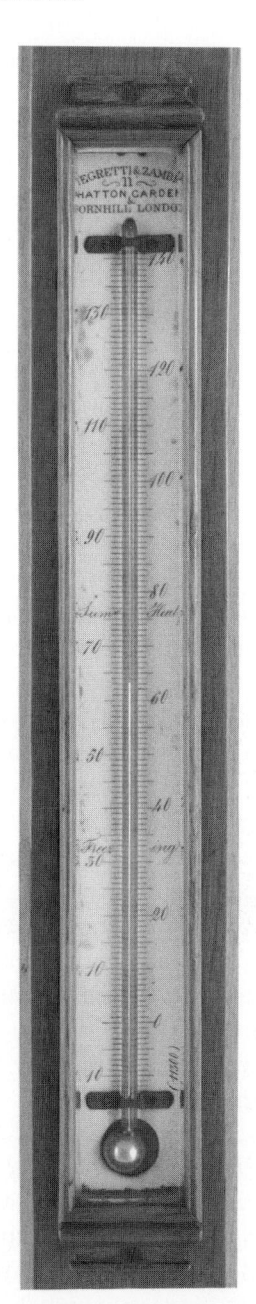

Figure 4.12 Thermometer scale of figure 4.11 showing fine text of early design.

Figure 4.13   Barometer scale of storm barometer in figure 4.11.

the 1870s. The thermometer has the later, bolder text created by transfers being fired into the glaze, and, if studied carefully, the case is slightly different: these items were not machine made but individually produced by hand to similar designs. Figure 4.15 shows the top scales and the brass screws that were commonly used to secure the wood frame around the scales. As with the marine or 'gun' barometer, brass screws and fittings were used throughout the construction. Although brass corrodes, the screws would not rust and disintegrate when positioned in extreme conditions near the sea. Many have survived for a hundred years and more in a very reasonable condition.

Figure 4.16 illustrates another variety of storm barometer, probably dating from around the 1860s, made by Negretti and Zambra, again in oak and with the same ceramic scales, but with a hinged door. This barometer is marked on the brass plate below the scales 'FB No. 23' and is certainly a 'Fisheries Barometer'; it is also marked 'Board of Trade' and is probably one of a batch made specifically for the Fisheries Board. Certainly, I have only seen one other like it (currently on display at the Meteorological Office in Exeter) so I do not think it is very common.

The case displays several differences, namely in the use of the hinged door, but also in the design of the thermometer box and scale and the urn-shaped cistern cover. Figure 4.16 shows where a plaque was fitted (usually 'On loan from the Meteorological Office') below the scales, and the setting key is incorrect, being the shank of an ordinary small cupboard key instead of a special brass-topped one. The barometer also has a remarkable feature that warranted preserving for the future when the barometer was restored. Visible in figure 4.17, showing the door open, is damage to the polish on the wooden case and blackening of the brass at the top of the scales. This is clearly damage caused by fishermen checking

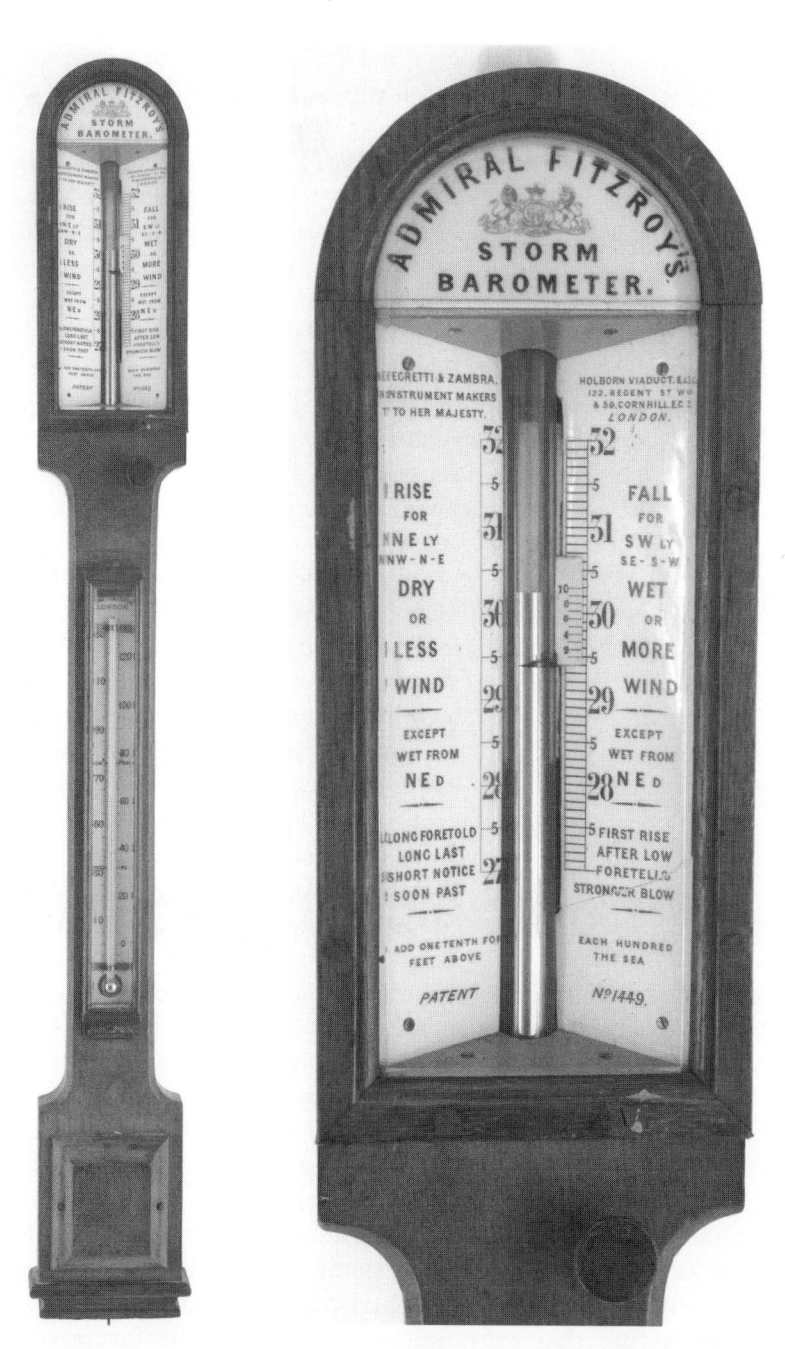

Figure 4.14 Storm barometer, number 1449, c.1870s, by Negretti and Zambra.

Figure 4.15 Barometer scale of storm barometer in figure 4.14.

FitzRoy Storm Barometers

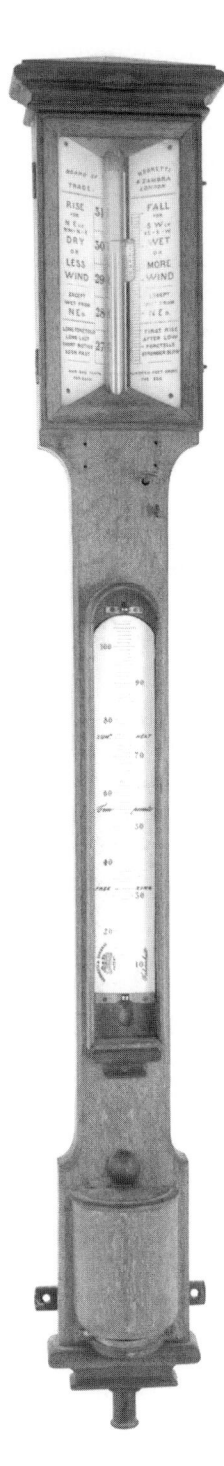

Figure 4.17 Barometer scale of 'Fisheries Barometer' in figure 4.16.

Figure 4.16 A 'Fisheries Barometer', number 23, c.1860, by Negretti and Zambra.

on the barometer during night time, opening the door and holding an oil lamp close to the scales to read the mercury. This evidence of use adds to the appeal of this fascinating storm barometer.

A late style of storm barometer is shown in figure 4.18, now only titled 'Admiral FitzRoy's Barometer'. Probably made to look like, but not actually by Negretti and Zambra, the barometer is similar in general style, but the thermometer case and scale resemble those found on much later barometers. The close-up illustration of the scales in figure 4.19 shows quite thick wooden edging, and the scales are in white glass. The instrument possibly dates from around 1910.

Apart from the storm barometers that FitzRoy promoted and Negretti and Zambra marketed, many other barometer makers sold similar items, although it is not possible to show all the rich variations of designs here. They all have FitzRoy's weather words or rules in common to a greater or lesser degree. Some would have been in direct competition with Negretti and Zambra's storm barometers; many others were in the style of the times and incorporated some of FitzRoy's weather rules. Figure 4.20 shows the top section of one such barometer; it is by another famous maker, Henry Hughes of 93 Fenchurch Street, London, where he worked after 1876. The scales are engraved ivory, typical of the time. Figure 4.21 is a heavy oak-cased barometer in the FitzRoy style; the close-up illustration in figure 4.22 shows the scale which has red and black text. Many makers as well as Negretti and Zambra used flat-top cases rather than the rounded design mostly promoted by Negretti and Zambra.

Fisherman's Aneroid Barometers

Also of significance for the preservation of life around the coasts were the thousands of aneroid barometers produced by Negretti and Zambra and by Dollond for the RNLI. Some of these were given as prizes, but the bulk were given or supplied at below cost price to fishing smacks and coastal vessels, so important was a barometer (and, indeed, still is) when sailing. The large mercury storm barometers supplied to fishing villages were only of use on land, and were not suitable for use at sea in comparatively small fishing boats. It was therefore suggested to the Committee of the RNLI by the chairman, Sir Edward Birkbeck, that they might begin to issue aneroid barometers to fishermen. In June 1882, therefore, Richard Lewis, on behalf of the RNLI Committee, sent out a letter to request the displaying of a circular (or poster) which explained the reasons why the RNLI had decided to supply aneroid barometers to fishing vessels. He wrote that: 'it is notorious that the masters of our small fishing craft hardly ever think of carrying with them an aneroid, and thus,

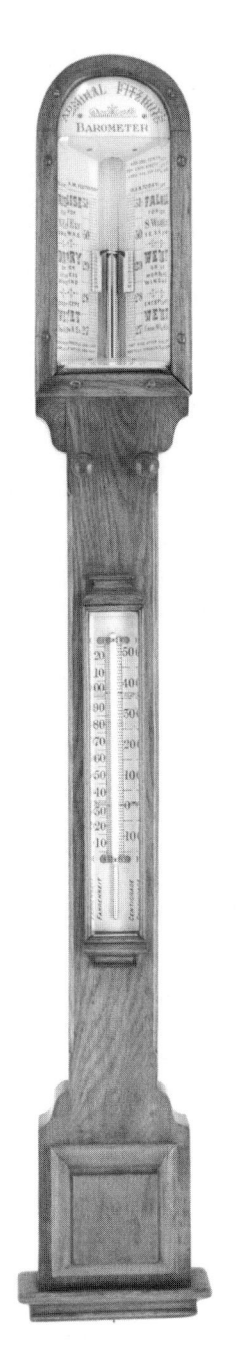

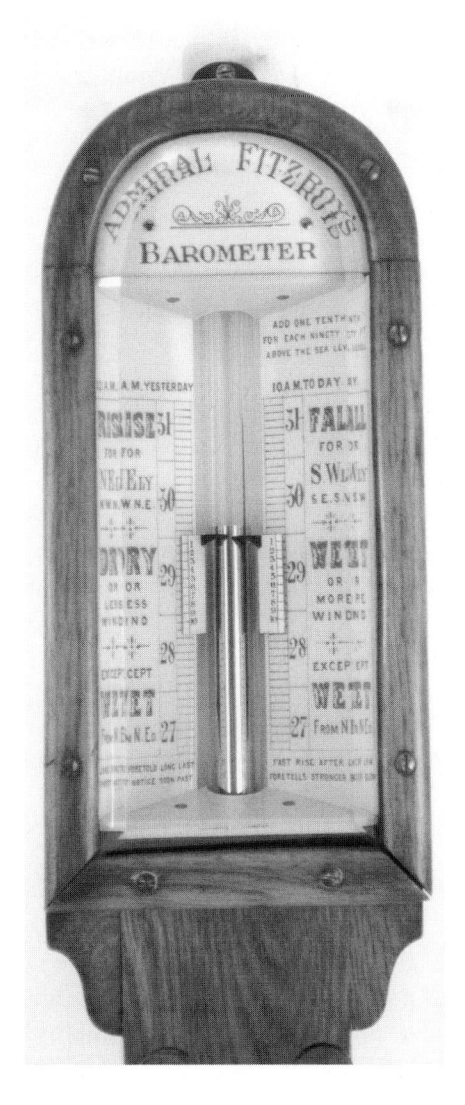

Figure 4.19 Barometer scale of 'Admiral FitzRoy's Barometer' in figure 4.18.

Figure 4.18 A very late 'Admiral FitzRoy's Barometer', c.1910.

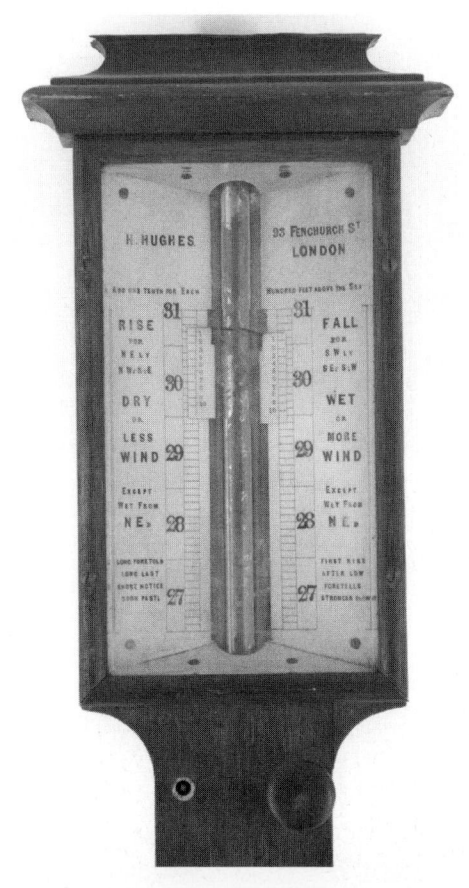

Figure 4.20 FitzRoy-inspired stick barometer scales by H. Hughes c.1885.

when in mid-ocean, they are without the most hopeful means of forecasting the disasters which too often overtake them when gales of wind suddenly spring up.' His letter confirmed that the aneroids could only be supplied to the owners or masters of fishing vessels and on no account to ordinary fishermen, sailors or the public.

The RNLI spared no effort in obtaining a good instrument and one that would not easily get out of order on board fishing smacks or require frequent repair. The RNLI then issued the following information to their lifeboat stations and other places:

The attention of the National Life-Boat Institution has been called to the great use an aneroid barometer would be to fishing boats, for indicating the approach of bad weather. It is a matter of experience that notice of a coming gale is often given by an aneroid – as it is by an ordinary mercurial barometer – some hours before the storm is actually felt. It seems reasonable, then, to hope that with such an instrument placed in the hands of intelligent fishermen, such men might be prevented from putting to sea when a gale is imminent, or be enabled to decide when it is prudent to run for a port: and thus some mitigation in the loss of life on board fishing vessels, through their being overtaken by gales of wind, might be looked for. Certainly fishermen, in common with other seafaring men, would be benefited by its use, and probably be able to avoid some of the dangers so often at present disastrous to them.

In this way, the RNLI invited applications for aneroid instruments at a

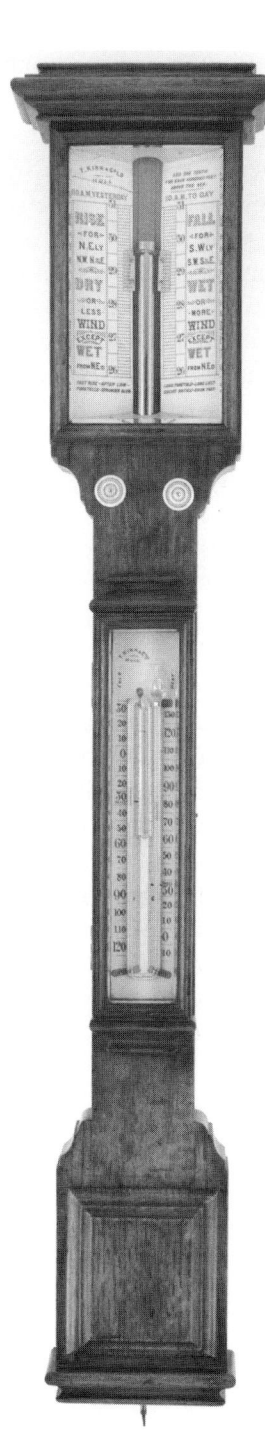

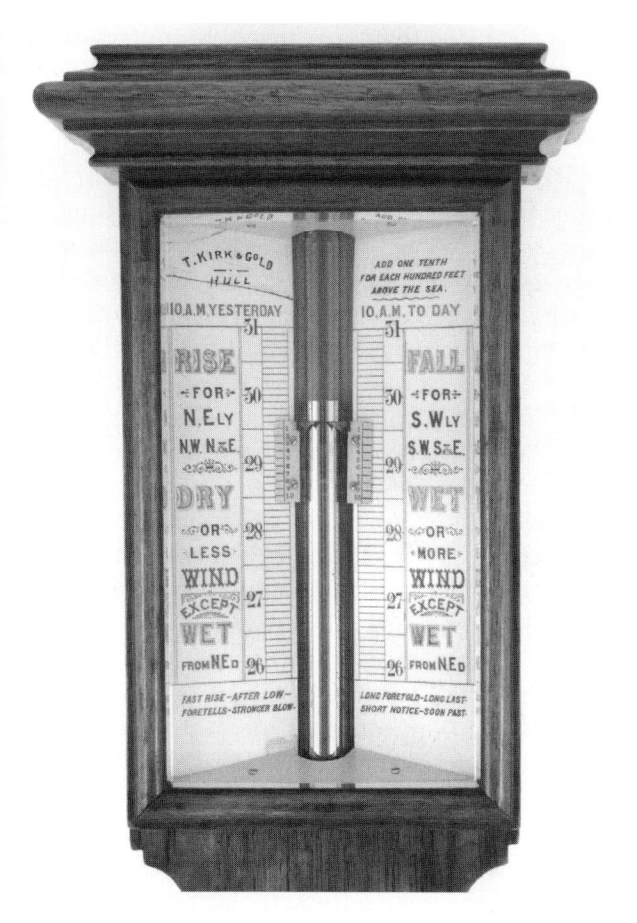

Figure 4.22 Close-up of the scale of the barometer shown in figure 4.21.

Figure 4.21 Heavy FitzRoy-type oak stick barometer of coastal design by T. Kirk and Co. Ltd, Hull.

great deal less than cost price. Many newspapers were sent the circular and many printed it in full, some in part, and with such publicity, requests came flooding in. Occasionally, an aneroid was supplied to a fisherman in his home, but this was rare and only by special permission. The RNLI charged 11s 6d for each instrument, although they cost approximately twice that sum to the Institution. On 23 June, the first order was received, and by the end of 1882, 883 had been ordered but only for supply to bona fide owners or masters of fishing vessels.

In October 1883, the secretary of the Institution sent another circular out to the lifeboat stations and fishing harbours: it repeated the previous offer of an aneroid, but now included coastal vessels under 100 tons burden, and commented that all seafaring men would benefit from its use. The first order for a coastal vessel aneroid (which was the same as the fisherman's in design) was received on 18 October 1883, and logs were kept of each order, either fisherman's or coastal vessels, separately until the demise of the scheme in the twentieth century.

Messrs Negretti and Zambra and Messrs Dollond and Co. (Dollond and Aitchison from 1927) received orders for aneroids alternately, so that by 19 December 1892 Negretti and Zambra had supplied 1,330 and Dollond and Co. 1,329 barometers. This pattern continued until after the First World War, although not always exactly equally, certainly without much difference. The last issue of this round was in 1925 when nine were issued, but only one each per year had been issued prior to that for three years. Tables 4.1 and 4.2 show the number issued to fishing vessels and coastal vessels respectively, and how the rate of issuing later declined: obviously the good barometer lasted well and did not need replacing often.

Table 4.1 Issue of aneroid barometers to fishing vessels

Date	Total issued from start
December 1882	883
April 1884	1,560
30 March 1885	1,845
4 August 1886	2,025
29 July 1887	2,127
27 March 1889	2,286
29 October 1891	2,557
14 December 1899	3,402
1 November 1923	4,750
16 January 1932	4,968

Table 4.2   Issue of aneroid barometers to coastal vessels

Date	Total issued from start
4 June 1884	111
7 February 1885	219
5 September 1885	262
28 January 1888	401
5 September 1895	637
27 December 1899	770
2 November 1925	979

Here the story of this benevolent service by the RNLI might have ended but, curiously, in March 1955, an order for 12 fisherman's aneroids was received from O. Comitti and Son Ltd of London at £4 each and were offered to bona fide fishermen at £2 5s each. Of these instruments, numbered 1–12, numbers 4, 8, 9, 10, 11 and 12 were returned for re-setting as readings were low and not consistent; number 8 was returned damaged. The quality of barometers after the Second World War was not apparently as good as in the nineteenth century, a fact that I have often discovered whilst restoring barometers of varying ages. Barometers numbered 13–24 were received in August 1956, costing £4 5s each; two dozen more (25–48) were ordered in April 1962, costing £4 15s 6d each, and the cost was to be increased to fishermen to £2 10s. Each barometer bears the trade mark of O. Comitti and Son, an anchor flanked by the letters O. and C.

Just how the numbering system of these fisherman's aneroids worked is still a mystery: perhaps Negretti and Zambra and Dollond arranged to number each barometer in numerical order and reserved the numbers over 10,000 for use on sales of the same barometers to the public and others. Certainly, they must have sold many thousands as numbers over 20,000 can be seen by each manufacturer. Negretti and Zambra continued to offer the fisherman's aneroid for sale in their later catalogues, although it had disappeared by 1931 in their E5 Engineering and Industrial Instruments catalogue; a very similar cased barometer is offered but not the traditional fisherman's aneroid. It does, however, appear – in traditional design as if from their catalogues of the 1880s – in what must have been their last barometer catalogue (ref no. b/17) of the early 1970s, priced at £18 and 5 new pence.

Figure 4.23 is a typical fisherman's aneroid barometer by Negretti and

Figure 4.23 Fisherman's aneroid barometer by Negretti and Zambra issued by the RNLI, c.1890.

Figure 4.24 Fisherman's aneroid barometer by Dollond of London issued by the RNLI, c.1902.

THE
"FISHERMAN'S" Aneroid Barometer.

"FISHERMAN'S" ANEROID BAROMETER, as supplied by Negretti & Zambra to the Royal National Lifeboat Institution.

This instrument is specially constructed to meet the decision of the Committee to supply a reliable Aneroid of a size available for use on board small fishing smacks and coasting vessels in which it would be impossible to use a mercurial Barometer.

No time, trouble or expense has been spared to obtain a really trustworthy instrument at a moderate cost.

Construction.—It is strongly mounted in a stout bronzed metal case, with thick plate-glass covering the dial, which is permanently enamelled on copper. The scale ranges from 26 to 31 inches, subdivided to half-tenths of an inch, and underneath is printed a condensed form of Admiral Fitzroy's rules for prognosticating the coming weather.

No. 252 ... ... £1 10 0

TRADE MARK.

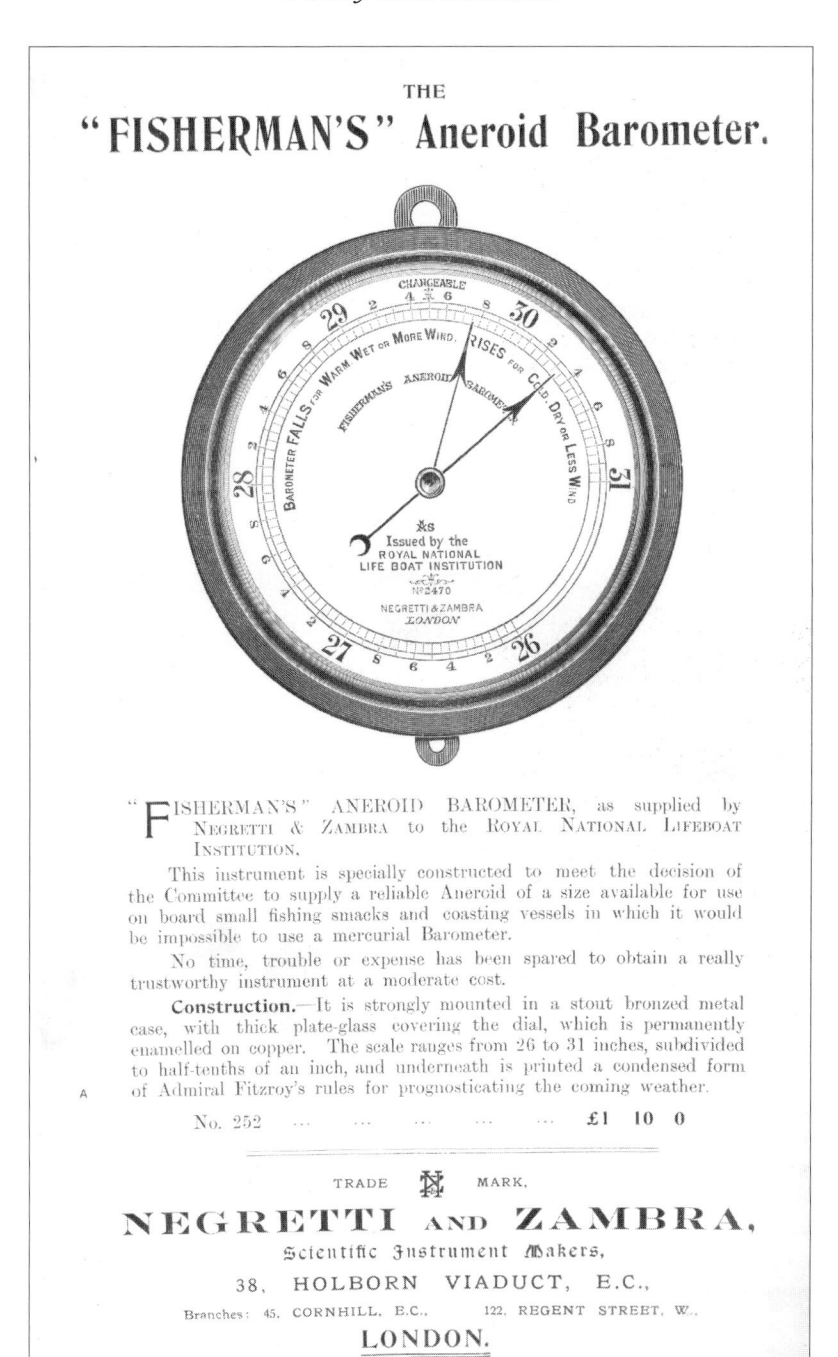

NEGRETTI AND ZAMBRA,
Scientific Instrument Makers,
38, HOLBORN VIADUCT, E.C.,
Branches: 45, CORNHILL. E.C.   122, REGENT STREET, W.,
LONDON.

Figure 4.25   Details of the fisherman's aneroid barometer from Negretti and Zambra's catalogue of 1913.

Marine Aneroid Barometer,
OF THE SHIPWRECKED MARINERS' SOCIETY.

"*To help Fishermen to Save their own Lives, and encourage them in Saving the Lives of others.*"

(From "Meteorological Office Instructions.")

IN the Aneroid Barometer, atmospherical pressure is measured by its effect in altering the shape of a small, hermetically sealed, metallic box, from which almost all the air has been withdrawn, and which is kept from collapsing by a spring. The top of the box is corrugated.

When the atmospherical pressure rises above the amount which was recorded when the Instrument was made, the top is forced inwards, and *vice versâ*, when pressure falls below that amount, the top is pulled outwards by the spring. These motions are transferred, by a system of levers and springs, to a hand which moves on a dial like that of a Wheel Barometer.

The Aneroid is very sensitive, and owing to its convenient size and portability, is especially suitable for Fishermen, Pilots, or Seafaring Persons employed in Boats or small Coasting Vessels, in which there is not space to suspend a Mercurial Barometer.

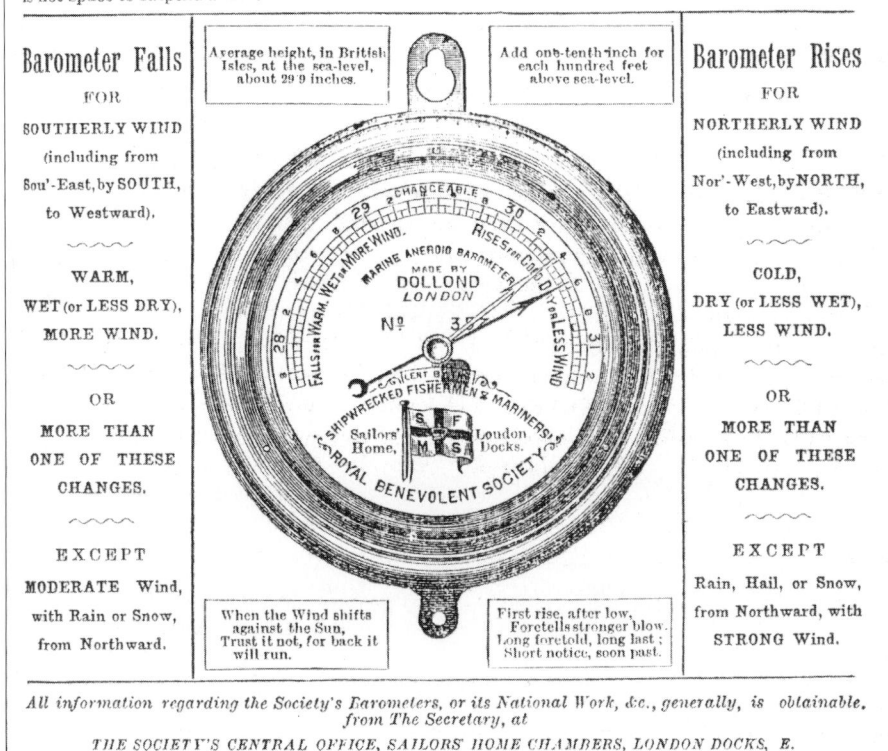

Barometer Falls	Average height, in British Isles, at the sea-level, about 29.9 inches.	Add one-tenth inch for each hundred feet above sea-level.	Barometer Rises
FOR SOUTHERLY WIND (including from Sou'-East, by SOUTH, to Westward).			FOR NORTHERLY WIND (including from Nor'-West, by NORTH, to Eastward).
WARM, WET (or LESS DRY), MORE WIND.			COLD, DRY (or LESS WET), LESS WIND.
OR MORE THAN ONE OF THESE CHANGES.			OR MORE THAN ONE OF THESE CHANGES.
EXCEPT MODERATE Wind, with Rain or Snow, from Northward.	When the Wind shifts against the Sun, Trust it not, for back it will run.	First rise, after low, Foretells stronger blow. Long foretold, long last; Short notice, soon past.	EXCEPT Rain, Hail, or Snow, from Northward, with STRONG Wind.

All information regarding the Society's Barometers, or its National Work, &c., generally, is obtainable from The Secretary, at
THE SOCIETY'S CENTRAL OFFICE, SAILORS' HOME CHAMBERS, LONDON DOCKS, E.

Figure 4.26 Details of the marine aneroid barometer of the Shipwrecked Fishermen and Mariners' Royal Benevolent Society, dated 1886.

Zambra of around 1890, with simple metal case and 5 inch diameter enamel dial, normally numbered though this one is missing its number. The similar barometer in figure 4.24 is numbered 3641, c.1902, by Dollond of London, again with an enamel dial. The dials are in red and black, with the numbers hand applied; both were issued by the RNLI.

Figure 4.25 is from Negretti and Zambra's 1913 catalogue, where the fisherman's aneroid is priced at £1 10s. Note the dial states 'as issued' by the RNLI: they were available for anyone to purchase, not just the RNLI. The Shipwrecked Fishermen and Mariners' Royal Benevolent Society, which was founded in 1839, also issued barometers to fishermen and coastal workers for the preservation of life, as well as awarding them as prizes; 43 of these barometers were given between 1882 and 1896 as rewards for saving life at sea. The society still flourishes today, giving support to fishermen, mariners and their families. These barometers may all have been made by Dollond, and have porcelain dials and wooden cases. Figure 4.26 gives details of the Society's 'marine aneroid barometer' from 1886.

By 1932, close to 200 mercury storm barometers and 6,000 aneroids had been issued by the RNLI in an effort to preserve life, and many others were erected around the British coast. The promotion of the aneroid no doubt also encouraged other mariners to buy a barometer if they could not get a 'subsidised' one. FitzRoy's contribution to weather forecasting amongst the poor fishing villages by barometer supply alone was considerable and his effect on the loss of life immeasurable.

5   FitzRoy's Barometer Legacy

As we have seen, as chief of the Meteorological Office Robert FitzRoy instigated the distribution of two types of barometer: marine barometers for use on board ships (see chapter 3) and storm barometers, as they were later called, for use by fishermen and others in ports and harbours around the coast (see chapter 4). Both types of barometer were also used to supply FitzRoy with data for his continuing investigations into the weather and for his pioneering work in weather forecasting. However, after FitzRoy's death, another type of barometer began to be associated with his name: this is the domestic barometer that is often described (somewhat misleadingly perhaps) as 'Admiral FitzRoy's Barometer', but which is clearly quite different from the two types of barometer that FitzRoy was familiar with.

In 1862, FitzRoy published *The Weather Book*, which was a success and went into a reprint. In his sixpenny *Barometer Manual*, a practical guide for ordinary users, such as fishermen, farmers and gardeners, he repeated his explanation of the functioning of the barometer and how to interpret its readings. Such was FitzRoy's influence on interpreting the barometer that after his death many barometers were made that incorporated his weather rules and indications. In this way, FitzRoy gave to barometer-making a long and lasting legacy.

Storm or coastal barometers, as we have seen, were produced with simple explanatory scales to help users to understand the rise and fall of the barometer according to wind direction, which can have considerable effect on the column of mercury. After FitzRoy's death, some large-cased domestic barometers began to appear which mostly used paper scales to incorporate some of FitzRoy's weather rules and were clearly inspired by his interpretation of weather indications. It is not known whether FitzRoy saw any of these paper-scaled domestic barometers, but they proliferated after his death and many thousands still survive in the UK. It is likely that the earliest types, dating from around 1865, are those simple, narrow-cased instruments that occasionally appear on the market but are often overlooked by dealers and collectors alike because of their simplicity and unassuming design. I rather like them.

Figure 5.1 shows one such item in an oak case with half paper scale and half polished wood, mounted with a boxwood thermometer engraved with the maker or retailer's name and with a simple, peardrop-shaped bulb tube. The top of the printed scale has the same scale words found on storm or coastal barometers, along with FitzRoy's rhyming words 'Long foretold long last, short notice soon past' and 'First rise after low foretells stronger blow.' However, below this, six further indications are printed under the heading 'Admiral FitzRoy's Remarks'. The single, mercury-level indicating pointer is short and points to the scale not the mercury itself; this type of pointer soon gave way to dual indicators as will be seen below.

'Admiral FitzRoy's Remarks' tend to vary from model to model, but invariably the words are taken from FitzRoy's *Weather Book*. As the scales were mostly printed paper, the printing plates were often used for many years and so the same paper scale will often be used on a different design of case. On the model illustrated in figure 5.1, the words on the left-hand side read:

> Where the state of the weather appears to disagree with any great change of the barometer, such change may be looked for with double force at no great distance off. Thus in May 1857 in and around London the mercury which had been lowering for some days began to rise with but $\frac{3}{4}$ inch rain (per rain gauge) whilst Reading in Berkshire, only 30 miles off, was visited by a storm so severe as almost to form an event in the annals of the town.

This statement is curious in two features: first, the date can vary, so that later barometers give the year as 1861 for the same text; and, secondly, the text appears to be on the earlier models, perhaps before 1875. Below this text, printed parallel with the tube, is the following weather law: 'A sudden fall of the barometer with westerly wind is generally followed by a violent storm from NW.N or NE.' This law is normally incorporated into the general rules on later barometers when the format of the scales became more standardised.

A similar type of barometer, but with a full paper scale and the thermometer mounted on the left, is illustrated in figure 5.2. It has a mahogany case and the scales are quite different from any I have seen before. The 'Remarks' are fairly similar, but the main feature is the table at the bottom of the printed scale, more easily seen in figure 5.3. This is 'Symons's Barometer Table', but without any instructions I have no idea how it works. If any reader recognises it, or has instructions surviving, it would certainly be interesting to know the history behind the table and how to use it.

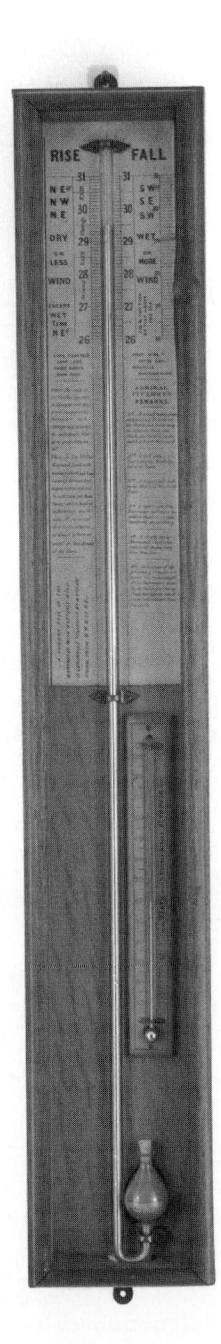

Figure 5.1 Early domestic FitzRoy barometer in oak case, c.1865.

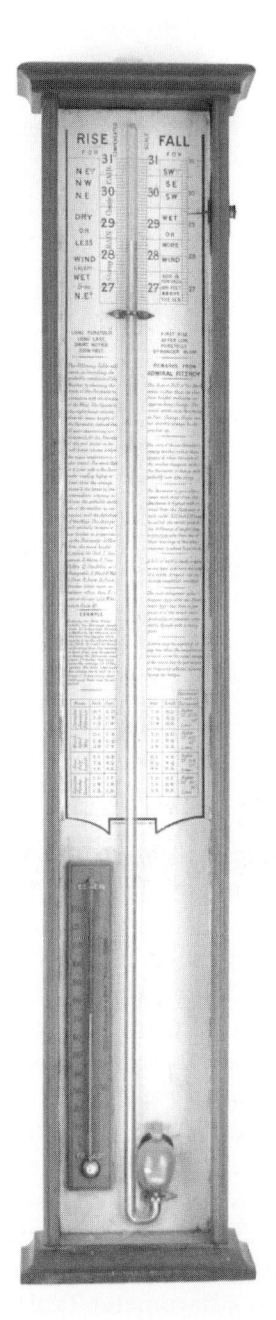

Figure 5.2 Mahogany-cased domestic FitzRoy barometer with 'Symons's Barometer Table', c.1865.

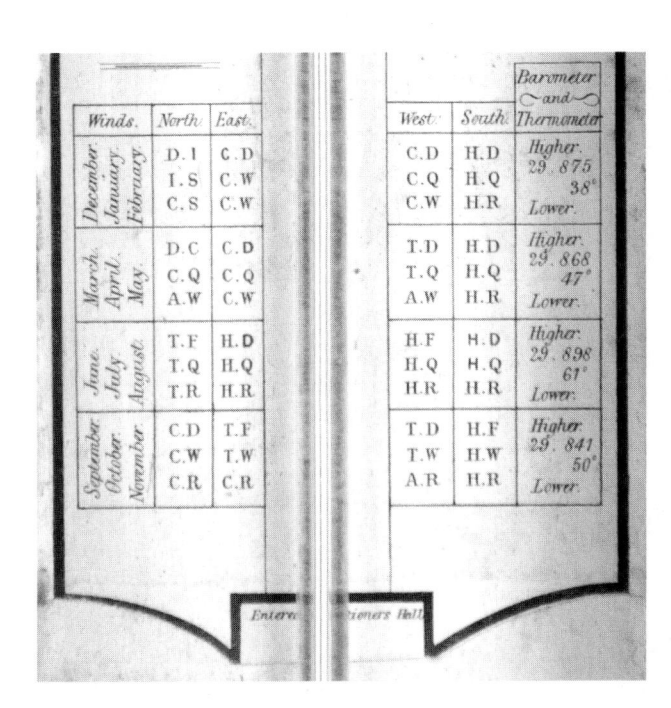

Figure 5.3 'Symons's Barometer Table' on the scale of the FitzRoy barometer shown in figure 5.2.

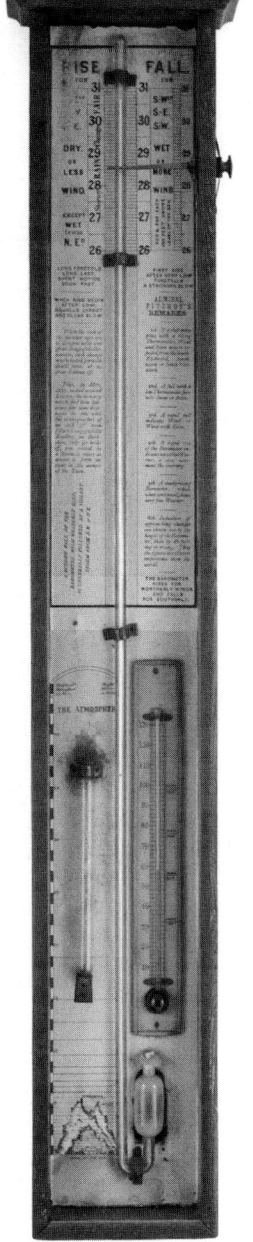

Figure 5.4 Early domestic FitzRoy barometer in oak case, c.1870.

Figure 5.4 shows a similar model, but with a full-length paper scale and the addition of a small storm glass or bottle (see below) mounted on the lower left-hand side above a scale indicating the height of mercury at different altitudes – a ready reckoner, as it were. This barometer is most likely later than that shown in figure 5.1, and is certainly not as good quality. None of the barometers discussed so far advertises itself as 'Admiral FitzRoy's Barometer', but they contain the same features as the models discussed below that have this title at the top of their paper scales.

The 'storm glass' (also known as a storm bottle, weather glass or camphor glass) in figure 5.4, mounted on the left of the main tube, towards the bottom of the case, seemingly in balance with the thermometer on the right of the tube, is a common feature on the domestic Admiral FitzRoy barometer. The popularity of the storm glass was probably due to FitzRoy's published interest in it. Whether or not FitzRoy had taken much notice of the storm glass before his appointment to the Meteorological Office is unknown; he does not record any readings whilst on board the *Beagle*. Some correspondence with Negretti and Zambra about the storm glass appears to be of an enquiring nature, so perhaps FitzRoy only began to take an interest in this item after his appointment to the Meteorological Office.

The storm glass is a hermetically sealed glass containing a liquid, although there is as much mystery about the composition of the liquid as about the reading of the glass. R. G. Raper from Chichester wrote to the Secretary of the Meteorological Office on 7 November 1863 enquiring about the composition of storm glasses: 'Sir, May I refer you to Admiral FitzRoy's book – 'The Weather Book' – page 441 on the camphor glass – and enquire in what proportions the composition there mentioned should be placed in the vial. I have long watched the Indications of a Glass of this kind and have endeavoured to get a correct one – but without success.' On the back of Raper's letter, as his draft reply, FitzRoy wrote: 'Say that our storm glasses are supplied by Messrs. N & Z Hatton Garden – and that we do not prepare the mixture for them at this office.'

As detailed by Negretti and Zambra in 1861, the storm glass is composed of a mixture of camphor, alcohol, potassium nitrate, ammonium chlorate and water, although the proportions of these ingredients and the manner of their preparation have been kept secret by all who make it. Negretti and Zambra warned their customers against inferior-quality storm glasses. Everyone who makes it has their own preferred method, and the mixture seems to improve with age; indeed, it is better stood for some months before using.

In many old storm glasses the liquid will have dried up due to the cork shrinking and the liquid evaporating, so it is normal to replace it with a new mixture. In some instructions, it is suggested that it is advantageous to allow air pressure in by inserting a hot needle through the wax seal and cork which helps to create changes in the crystals formed, but most instructions are for a corked and sealed unit. As the mixture undoubtedly reacts to temperature, storm glasses are best positioned outside, although this is impossible, of course, for one mounted on a barometer. (The opinion that barometers should be positioned on an outside wall may well have its roots in the FitzRoy barometer with the storm glass mounted on it being thought better in such a position; the myth is still alive amongst many people, although any reason is long forgotten.)

In his *Weather Book*, FitzRoy mentions storm glasses twice: he suggests that they 'contribute indications of small disturbances of air such as squalls or thunder storms on land ' (p. 209) and that 'camphor glasses' show 'the degree of electrical tension' and 'indicate unfailingly the relative presence of electricity':

> Much polarity, (plus or vitreous) electricity, is shown by crystallisation of the camphor, in leaf-like, spiky shoots.
>
> Minus (resinous or negative) indications, subsiding or melting camphor – falling rain or snow, and the look of the sky, assure one of a lessened tension. Increase of either characteristic implies action, or alteration, in the upper air, or at a distance horizontally, but within influential causation. Camphor glasses, in proper positions and duly attended, are most useful to a skilled eye and quick perception. (*The Weather Book*, p. 232)

In the Appendix to his *Weather Book* (pp. 443–5), FitzRoy comments on the 'peculiar effects' he has often noticed in storm glasses which do not seem to be caused by pressure, or solely by temperature, dryness or moisture, but that these alterations occur with electrical changes in the atmosphere. FitzRoy states that it has 'lately' been 'fairly demonstrated' that:

> if fixed undisturbed, in free air, not exposed to radiation, fire or sun, but in the ordinary light of a well-ventilated room, or preferably in the outer air, the chemical in a so-called storm glass varies in character with the direction of the wind – not its force, specially, though it may so vary (in appearance only) from another cause, electrical tension.

FitzRoy goes on to explain the various changes that can be seen in the storm glass 'if closely, even microscopically watched'. As northerly or polar air approaches, 'fir, yew or fern leaves – or like hoar frost – or even large but delicate crystallisations' can be seen to grow in the glass. If the wind comes from the opposite direction, then the hard features of the crystals gradually soften and vanish. During southerly winds, the mixture sinks downward in the glass vial, becoming shapeless 'like melting white sugar'. During continued northerly winds, 'the crystallisations are beautiful' but the least disturbance unsettles the crystals. 'Stars' in the mixture 'are more or less numerous and the liquid is dull, or less clear' during easterly winds. In westerly wind, 'the liquid is clear, and the crystallisation well defined, without loose stars'. When there is positive electricity, then 'hard or crisp features are visible below, above or at the top of the liquid'. When only 'soft, melting, sugary substance is seen', the atmospheric current is southerly with negative electricity. FitzRoy believed that temperature affected the mixture greatly but not solely, 'as many comparisons of winter and summer' had proved. He continues:

> A confused appearance of the mixture, with flaky spots, or stars, in motion, and less clearness of the liquid, indicates south-easterly wind, probably strong – to a gale.
> Clearness of the liquid, with more or less perfect crystallisations, accompanies a combination, or contest, of the main currents, by the west, and very remarkable these differences are – the results of these air currents acting on each other from eastward, or entirely from an opposite direction, the west.

FitzRoy suggests that the liquid should be gently shaken two or three times a year and the glass wiped clean. Of course, this advice is of no use for a storm glass mounted on a barometer (unless it is removed). According to FitzRoy, repeated trials with a 'delicate' galvanometer to measure electrical tension in the air 'have proved these facts', which, he states, 'are now found useful for aiding, with the barometer and thermometers, in forecasting the weather'. FitzRoy perhaps summed up his views on the storm glass when he wrote: 'I think [the storm glass] is affected by the electrical condition of atmosphere – and useful as an indicator – in connection with other instruments – but not alone' (letter BJ7/331 8918, undated and unaddressed).

References to storm glasses in catalogues are scarce, although Negretti and Zambra included three different examples on page 147 of their encyclopaedic catalogue of around 1880 (figure 5.5), ranging from a simple

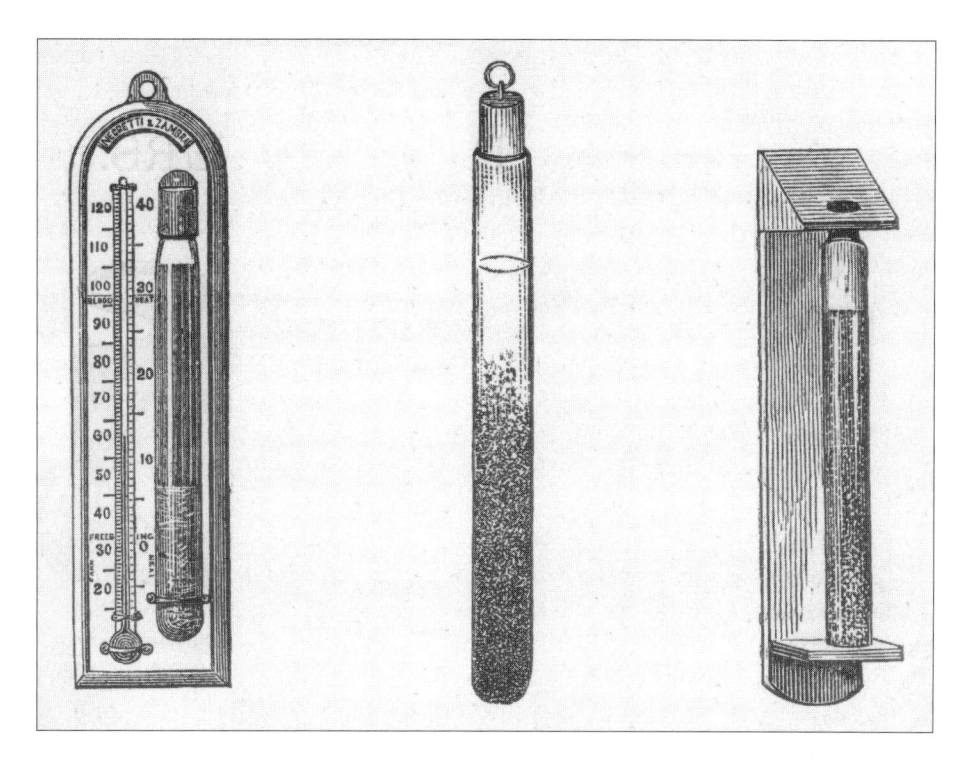

Figure 5.5 Storm glasses as offered in Negretti and Zambra's catalogue, c.1880. *Centre:* basic storm glass; *left:* storm glass mounted on a boxwood scale with thermometer; *right:* storm glass mounted on a window bracket.

glass container priced at 4s 6d to one mounted on a window bracket for £1 15s. They are listed as being useful 'for prognosticating changes in the weather, by sea or land, particularly high winds, storms, or tempests'. At the bottom of the catalogue page, however, Negretti and Zambra appear to cover themselves by drawing attention to two pamphlets written by Charles Tomlinson of King's College, London, in which 'it would appear that the changes observed in the Storm Glass are due solely to variations of light and heat.'

Clearly, Negretti and Zambra did not guarantee the prophetic virtues of the storm glass but were prepared to sell them. It may even have been Negretti and Zambra who encouraged the sale of these items by their marketing methods, as was the case with FitzRoy's 'coastal barometers' which they renamed 'storm barometers' much to the annoyance of FitzRoy's temporary replacement at the Meteorological Office, T. H. Babington (see chapter 4). However, even Negretti and Zambra later removed storm glasses from their list of instruments for sale as none is shown in or after their 1901 catalogue.

Nevertheless, since these curiosities had been brought into the weather-forecasting arena by FitzRoy, it is not surprising that when the domestic FitzRoy barometer began appearing soon after his death in 1865, to satisfy what must have been ready public demand, the storm glass was soon added to the decorative array of these types of barometer. The earliest domestic FitzRoy barometers (as seen in figure 5.1) do not normally have a storm glass, but certainly by the early 1870s it was a common feature, with perhaps only the most basic economy models lacking one.

The style of storm glass most commonly found, held in simple brass mounts, can be seen in figure 5.6. The storm glass in figure 5.7 has a more elaborate brass mount normally found only on more expensive, better-quality FitzRoy barometers. Storm glasses were also produced as separate 'instruments', and one can come across a wide variety of styles from simple 'cottage' types on very cheap wooden mounts with printed instructions to large ceramic-backed ones for public display. These items were certainly around up until the Second World War, as a number of people have told me that they remember seeing them outside opticians' shops, but that they were removed during the war for fear of damage and only a few were replaced; some are still in private collections and very occasionally one such item comes up for sale or in need of repair. As domestic FitzRoy barometers were made with them, later reproductions have included them and so interest in the storm glass continues.

From patent number 652, dated 20 February 1873, by a George Alexander Simmons of Great Saffron Hill, Middlesex (a known area of barometer manufacture), we find described a method of plugging barometers that have a bent glass tube such as commonly used in domestic Admiral FitzRoy barometers. The patent describes a device that can be used to raise and lower a plug to stop the movement of mercury either by a rack-and-pinion device, operated by turning a small knob mounted on the front edge of the case, or by a thumb piece at the rear of the barometer, which lifts a lever connected inside the case of the barometer. Without this device, the only way to stop the movement of mercury would be to remove the glass front and manually plug the tube, as would have to be done with barometers such as the one shown in figure 5.1. Figure 5.8 shows the plugging device operated manually by a thumb piece at the rear of the barometer, which proves that instruments with this feature date from after 1873 at least. Very few of the rack-and-pinion-operated plugging devices appear on the market, although I have seen some. It is slightly more common to see the rack-and-pinion mechanism on sympiesometers for the same reasons of transportation.

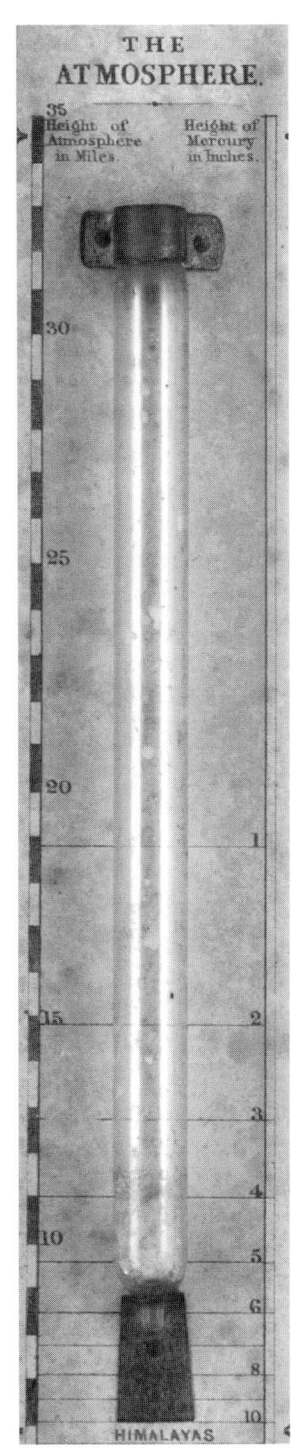

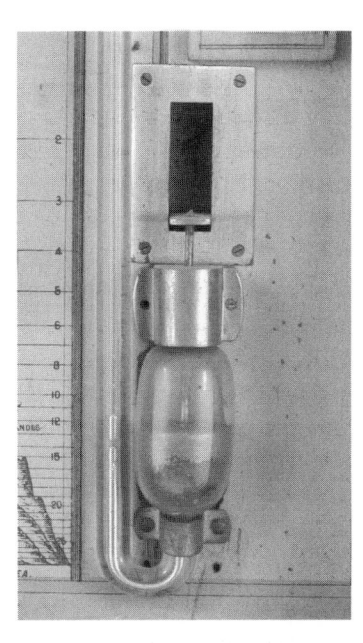

Figure 5.8 Detail of the 1873 patented plugging device on a domestic FitzRoy barometer. The plugging device is lifted manually from the rear, as opposed to the rack-and-pinion operated by a key which would be inserted into the case to the right-hand side of the device, though this latter type is rarely seen.

Figure 5.7 Storm glass with more elaborate brass mounts found on better-quality domestic FitzRoy barometers.

Figure 5.6 Style of storm glass commonly found on domestic FitzRoy barometers, held in simple brass mounts.

Figure 5.9 shows an early domestic FitzRoy barometer with carved top and the patented plugging device. Ones that have the rack-and-pinion method of moving the plug are of this style with the knob from the indicator pointers being used to turn the pinion which is set into the side of the case at a point approximately level with the plugging device on the right-hand side. The extra friction created by this design probably made the practical use of it difficult: there was more force needed to move it up and down, and the metalwork needed to be quite substantial to work well; hence the discontinuation of it in favour of the simple manually operated system shown in figure 5.8.

The barometer scale is printed on paper, but this time with the words 'Barometer by the Late Admiral FitzRoy' as a banner at the top (figure 5.10). Just how long after his death barometers might have been made and marketed describing FitzRoy as 'the Late Admiral FitzRoy' can only be conjecture, but perhaps we can date it to within ten years of his death. The presence of the transporting device would make it later than 1873. The barometer also has a silvered, engraved thermometer scale with flattened bulb (i.e. the bulb is not cylindrical in section but thin and wide), and the storm glass has an engraved boxwood scale with the weather indications 'Set Fair, Fair, Change, Rain, Stormy'. These features indicate a more expensive model (see figure 5.11).

Figure 5.12 shows the simple 'Oxford-framed' FitzRoy barometer (probably so named after the 'Oxford corner', a design of border used by printers where lines cross at right angles and project beyond the crossing point; this type of frame was very popular for prints from the 1870s). I always considered this later in design, but the paper scale (see figure 5.13) is exactly the same as that in the barometer shown in figure 5.9, so it is likely to date from the same time (although, once a printing plate had been made, it is quite possible that it was used for many years). The retailer's name, 'E. Rossiter Teignmouth', is included on a small boxwood plaque at the bottom of the case on the left (see figure 5.14).

It has long been my belief that the domestic FitzRoy barometer, with full-length glass front and paper scales, was the first mass-produced barometer in Britain. Perhaps there were other cheaper barometers, but in general a barometer had an engraved or porcelain scale and a cabinet-built case. The domestic-style FitzRoy allowed simple cases, cheap scales and was quite an intriguing wall hanging. With thermometer, storm glass and various texts to read, it was and still is an article of curiosity which is collectable in its own right.

Accuracy, however, is often not the strong point of the domestic FitzRoy barometer, which is another reason to think that they were frequently

Figure 5.10 Top of barometer in figure 5.9, showing banner heading.

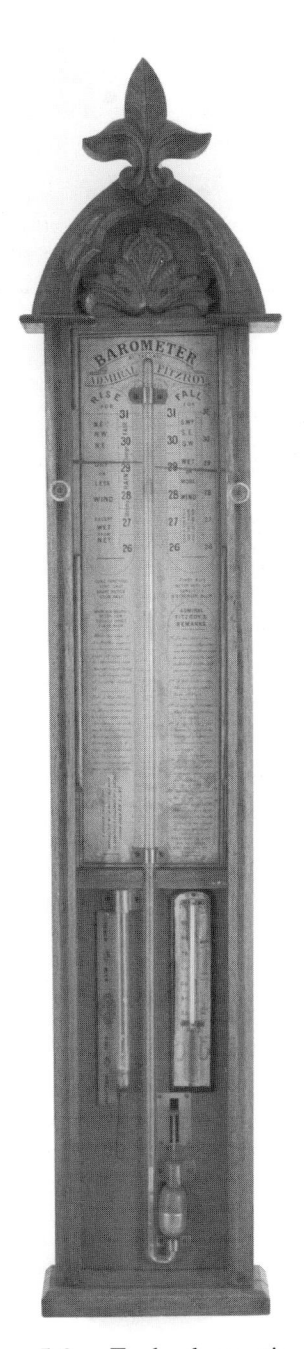

Figure 5.9 Early domestic carved FitzRoy barometer in oak case, c.1875.

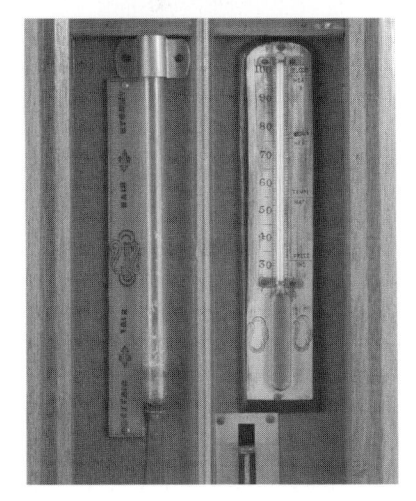

Figure 5.11 Storm glass and thermometer of barometer in figure 5.9.

Figure 5.13 Top of barometer in figure 5.12.

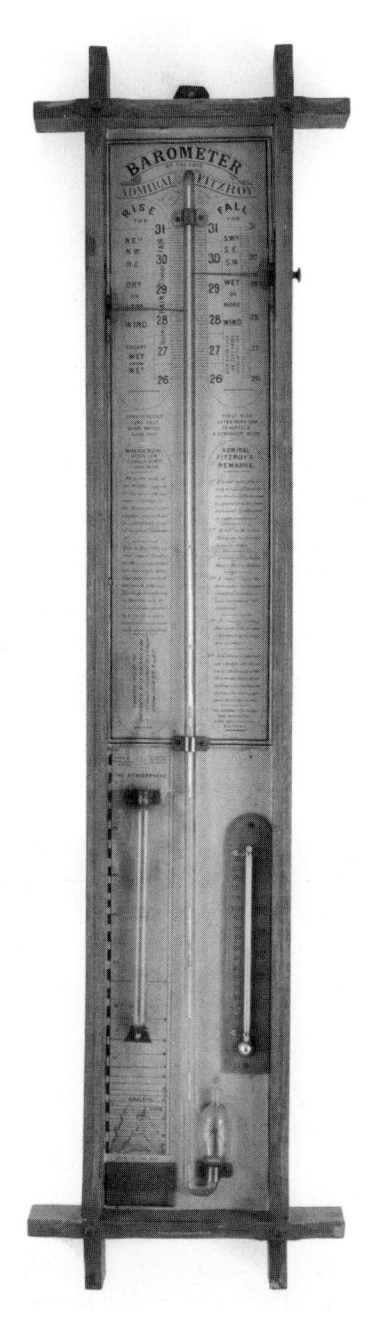

Figure 5.12 Oxford-framed domestic FitzRoy barometer in oak case, c.1875.

Figure 5.14 Base of barometer in figure 5.12, showing retailer's plaque.

mass produced. If you measure from the 30-inch mark on the scale down to just above the widening part of the bulb (where the level of mercury normally is in the bulb), this should be 30 inches, but often it is not, and errors of up to an inch are not uncommon. This probably results from a poor understanding of barometer-making and not positioning the scale in the correct place or not making the barometer tube to the correct length. If these barometers were mass produced by makers buying in the components, then a lack of knowledge of the technical aspects required to make a barometer most probably means that they were never built accurately in the first place.

To be generous to the makers, it is possible that they were made for different altitudes, but this is not an argument that carries much weight with me. They were simply built by craftsmen who were less skilled than traditional instrument-makers: they were made cheap and sold cheap. Many were made to higher standards, but the bulk was produced for the mass market. Unlike clocks, when you notice if the time is wrong, most people had only one barometer and, without comparative weather forecasts from the radio or television, few would notice any difference in reading. And, after all, they were not sold as scientific items, but as decorative barometers for the home – they were part of history, part of FitzRoy's legacy.

Figure 5.15 shows a domestic FitzRoy barometer with a carved top. The paper scales of many of these are sealed with shellac and have a cream-coloured, shiny finish which does help to protect them. The scales frequently survive far better than the plain paper scales seen on the Oxford-framed variety. The barometer has, as is common, two indicating pointers which are designed to be used for two days' history of barometer readings: one yesterday and one today, so that on the following day you have three readings to compare before altering the pointer with the small knob that turns the rack-and-pinion device to move the pointer. These 'keys' frequently get lost (as has happened to one in the illustration here) as they are only a push fit; they are often topped with wood, but can sometimes be topped with bone. This particular barometer has a later repair to the tube which can be seen by the brass box that a repairer has made to cover what I suspect will be a goodly amount of glue or resin to try to keep a broken tube intact. The atmosphere scale is coloured deep blue near sea level, fading to light blue at the higher reaches of the atmosphere; on a few, the height of the highest balloon flight is recorded. This may refer to the near-fatal flight by the famous balloonists Henry Coxwell and James Glaisher in 1862 when they reached a height of 30,000 feet (Glaisher estimated it at 37,000 feet).

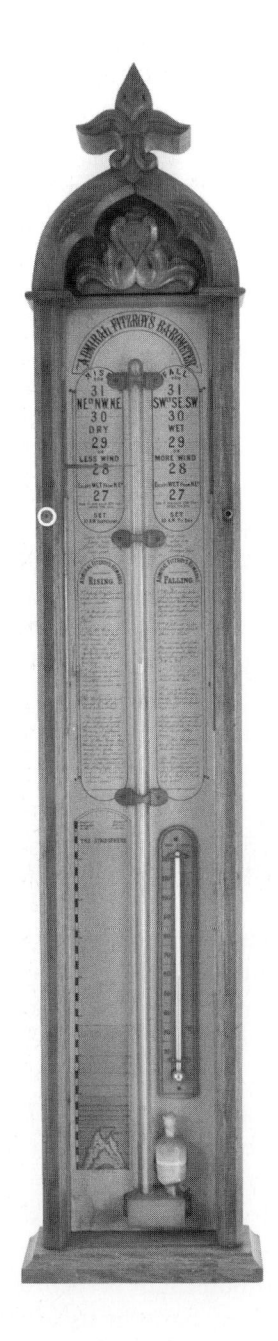

Figure 5.16 illustrates a similar model, this time with a storm glass, though the carving above the paper scales has never been fitted. The tube is a thick replacement with an especially long and wide bulb to cover a previous poor repair when a wrong-sized tube was fitted and the case carved to fit which left a gaping void where the original tube had been. Figure 5.17 is a domestic FitzRoy barometer of similar form but with an ivory thermometer scale mounted on a piece of black polished wood. Figures 5.15–5.17 all have the same design of paper scale and may have been made by the same manufacturer at different times. It seems that once a printing block had been made, many copies were produced over many years before it was discarded.

Figure 5.18 is probably of a later design; the scales are of the gothic style common to many of the simpler, Oxford-framed types. There are two brass plaques just below the scales which read 'FitzRoy's Barometer'. The design of the carving is another variation, and the carving below the crest is in front of the glass, not behind as many are. The carving is more detailed, with textured branches and leaves centred by a pseudo-heraldic shield. The crystals in the storm glass are, as often, dried up.

Figure 5.19 is more ornate than many of this type. The standard design case in American walnut includes a printed description of the storm glass indications on the small plaque beneath the storm glass; it also proclaims itself a 'Prize Medal Barometer' at the top of the scale. This probably means that this style of barometer was awarded a prize at an exhibition; there were many prizes given to barometer makers and they were not slow

Figure 5.15   Domestic carved FitzRoy barometer in oak case, c.1885, without storm glass ever having been fitted.

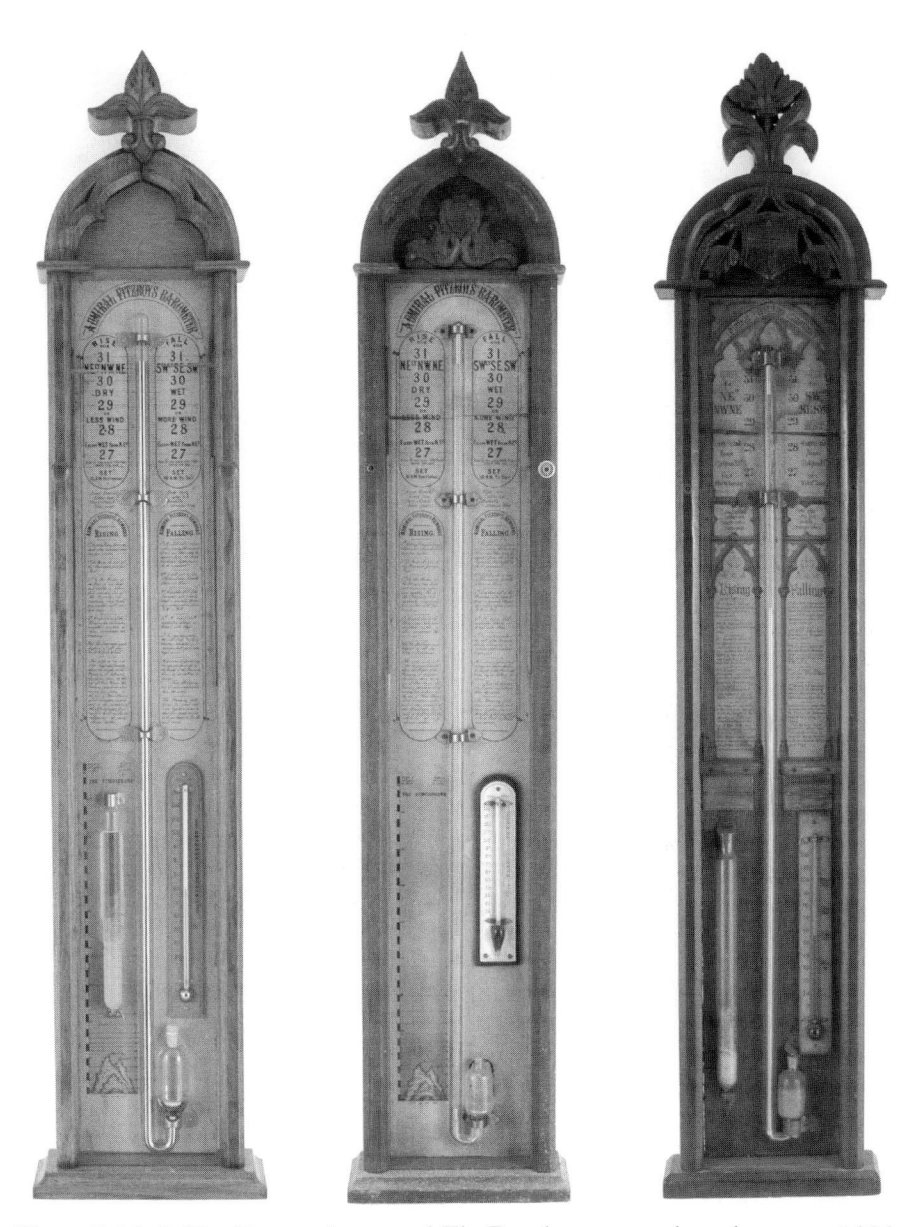

Figure 5.16 (left)   Domestic carved FitzRoy barometer in oak case, c.1890, without carving above the tube.

Figure 5.17 (centre)   Domestic carved FitzRoy barometer in oak case, c.1880, with ivory thermometer scale but without storm glass ever having been fitted.

Figure 5.18 (right)   Domestic carved FitzRoy barometer in dark oak case, c.1895, with gothic-style charts.

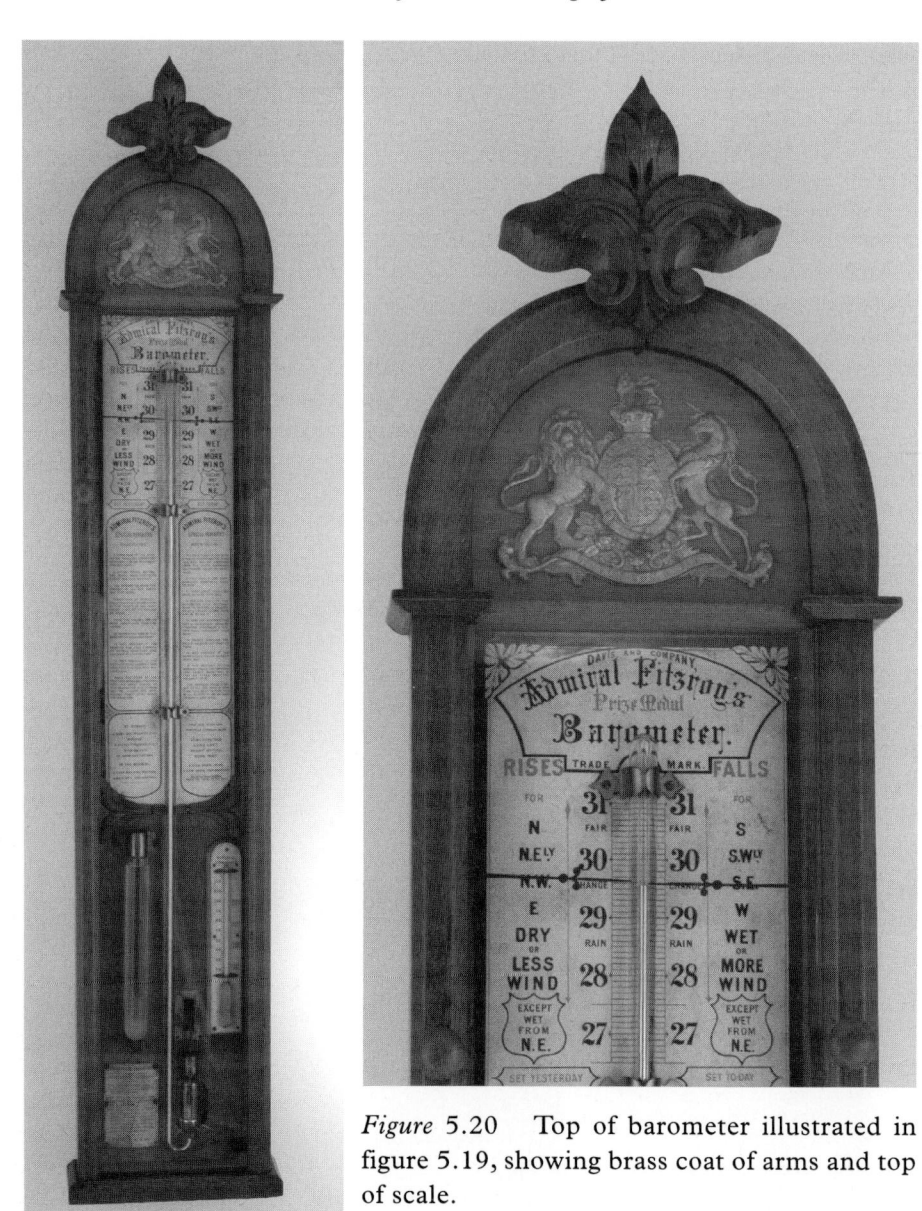

Figure 5.20 Top of barometer illustrated in figure 5.19, showing brass coat of arms and top of scale.

Figure 5.19 Domestic carved FitzRoy barometer in oak case, c.1885, by Joseph Davis and Co.

to advertise the fact. The ivory scale thermometer with flat bulb tube is engraved by the maker 'Davis and Co. of London', a company well known for making FitzRoy barometers. The most unusual part of this barometer is the brass Royal Coat of Arms mounted above the paper scales under the glass (see figure 5.20). This was most likely put on Davis and Co. barometers because they were working at the Royal Polytechnic Institute and saw it as good publicity, and not (as the owner had hoped) because the barometer had once been owned by a member of the royal family. Joseph Davis and Co. were more famously known for their 'Royal Polytechnic' barometers (see below) which invariably have the Royal Coat of Arms and Prince of Wales's feathers printed on their paper scales. The scales of this FitzRoy barometer are refreshingly distinct from the plethora of similar styles one comes across.

Figure 5.21 shows a very heavily carved domestic FitzRoy barometer with the patent plugging device, twin indicators, half paper scales and ivory-backed thermometer scale. Note also the more elaborate brass cover to the top of the storm glass. The paper scale is one that was produced quite late and possibly dates from the early twentieth century. It also appears on early barometers that have had replacement scales fitted due to restoration. The quality and appearance of this scale design is not one that I have ever really liked. Having inherited a few unused copies of this scale, I tend to think of them as modern. They were probably reproduced unchanged for many years and could still have been available in the 1960s.

The domestic FitzRoy barometer illustrated in figure 5.22 is in American walnut and has scales of very similar design to the previous barometer (possibly replaced at some time). It has a mirror where there should be a clock, a tall crested pediment, spring-type indicators and a patent plugging device. Figure 5.23 shows a rather simply carved (it could even be described as crude) pediment barometer. The scales are in very fine condition, but the tube has come from another instrument and the indicators have been adapted or badly repaired. The thermometer is engraved 'London and Universal Supplies Stores, 5 Victoria Avenue, Bishopsgate St, EC'. The design of the case differs from some FitzRoy barometers in that the front edge of the sides of the case is not moulded or bevelled but square and with a single groove down the centre, which would have been a simple way of cutting the cost of case-making. The same simple case design can be seen in figure 5.24, although the carved pediment is more standard and of better quality. Significant variations to this type of barometer, which were obviously made in their thousands, are rare, and dealers could probably say that they had seen them all after the first hundred or so.

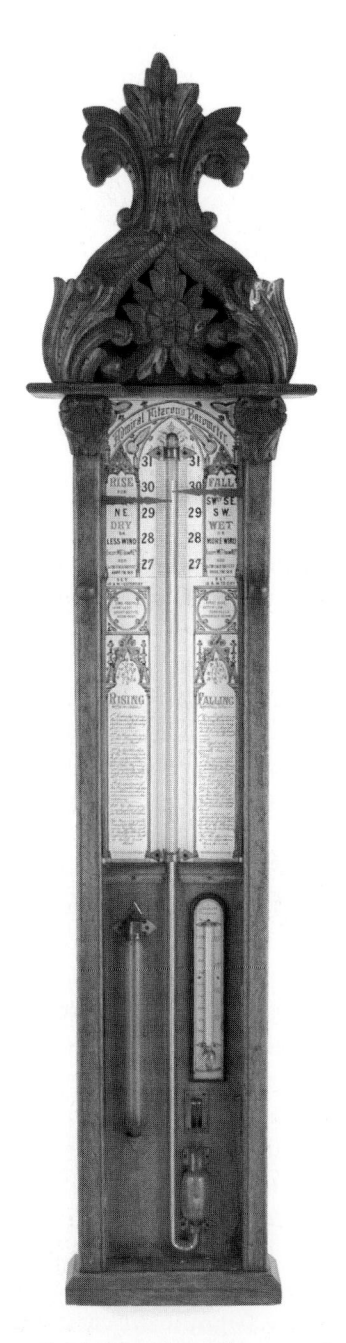

Figure 5.21 Very heavily carved oak-cased domestic FitzRoy barometer with replacement paper scale, c.1885.

Figure 5.22 Crested, thin-cased domestic FitzRoy barometer, c.1900.

Figure 5.23 Crudely carved pediment, thin-cased domestic FitzRoy barometer in American walnut case, c.1895.

Figure 5.24 Standard carved pediment, thin-cased domestic FitzRoy barometer in oak case, c.1895.

However, whilst the barometer shown in figure 5.25 at first looks like a standard, run-of-the-mill, Oxford-framed domestic FitzRoy barometer, closer inspection of a small, shield-shaped mark in the lower right-hand corner shows the patented transport device of 'WHM' (see figure 5.26). I have not discovered the details of this patent, but it is clear that the barometer tube is a thin stick tube with boxwood reservoir, and a key can be used to wind the transport screw seen protruding from the base of the case in order to make the barometer safer to transport. Finding uncommon designs like this one always adds to the enjoyment of handling FitzRoy barometers. The thermometer is engraved 'Squire and Son, Bideford', clearly a retailer and not a maker.

Figure 5.27 shows the archetypal domestic FitzRoy barometer, but with different paper scales. It also shows the false square-headed nails that are used on the intersecting corners of the Oxford frame to make it look as if they are holding the frame together; this is a very common feature on many of these designs. Figure 5.28 is a rather poor example of a domestic FitzRoy barometer with late Edwardian paper scales (probably replacements), an unsuitable replacement wood platform top of later making, a wire holding the storm bottle in place and the tube propped up with a cork; the wood has been scraped down with a blunt knife or similar and varnished. Sadly, it is all too common to come across specimens such as this in antique and collectors' shops. Many are the unfortunate casualties of keen enthusiasts who do not have the knowledge, eye, ability or desire to understand how to restore them to some degree of authenticity, although those that have survived intact are perhaps made all the more collectable by this sad fact. The thermometer and indicators are original though.

A style of barometer not often seen is illustrated in figure 5.29. It has a highly decorative, colourful paper scale complete with two ship scenes, depicting calm and storm, encircled by FitzRoy's weather rhymes (see figure 5.30). Above the storm glass and thermometer are two diagrams showing the solar system and the Earth (figure 5.31). The printing is in many colours, including green and yellow/gold on the border, orange, reds and browns; it is simply emblazoned with artwork. The case is in American walnut and the pediment houses a clock. I suspect that barometers like this one were the sole production of one company as few seem to appear on the market; those that do are often in good condition. As can be seen from the small selection illustrated so far in this chapter, the domestic FitzRoy barometer has been made for a considerable period of time and in many variations on a similar theme. They all have FitzRoy's weather words or rules in common and justifiably earn a place as FitzRoy barometers of a domestic character.

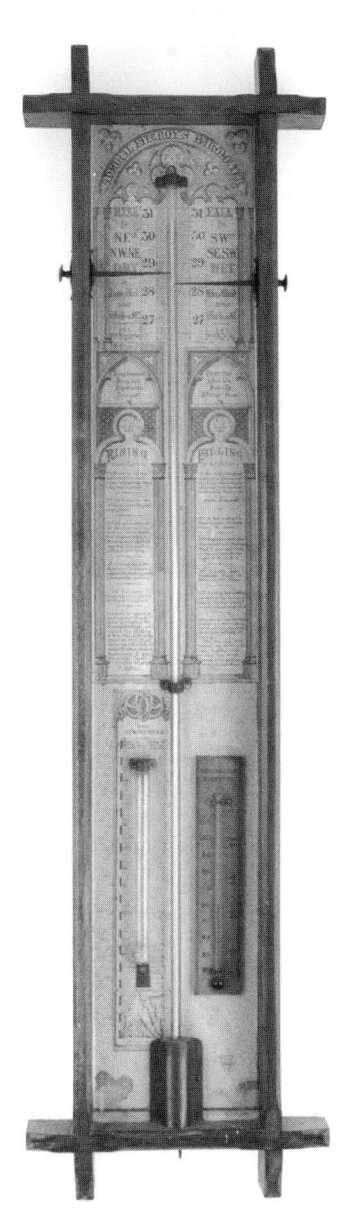

Figure 5.26 Transporting screw, tube cover and patent mark of barometer shown in figure 5.25.

Figure 5.25 Oxford-framed domestic FitzRoy barometer in oak case, with gothic charts and patent tube, c.1895.

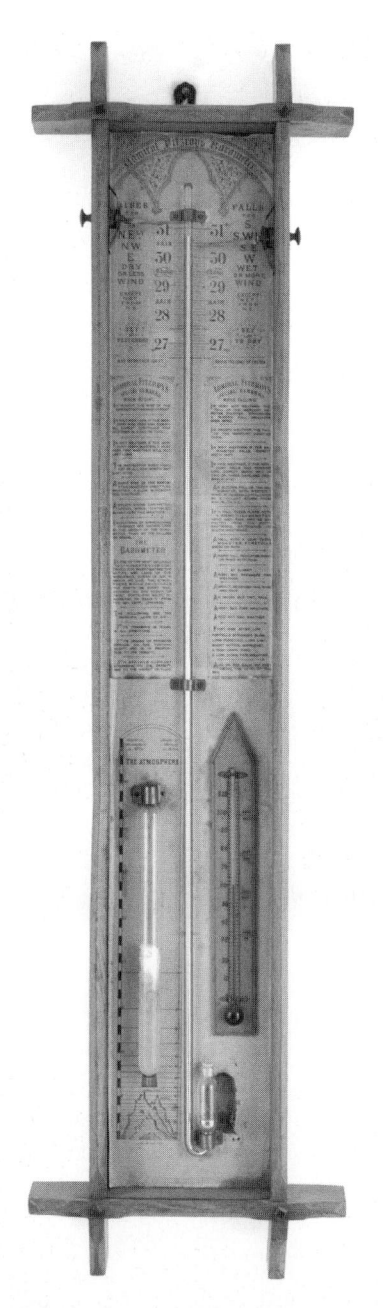

Figure 5.27 Oxford-framed domestic FitzRoy barometer in oak case, with mono (black) printed charts, c.1885.

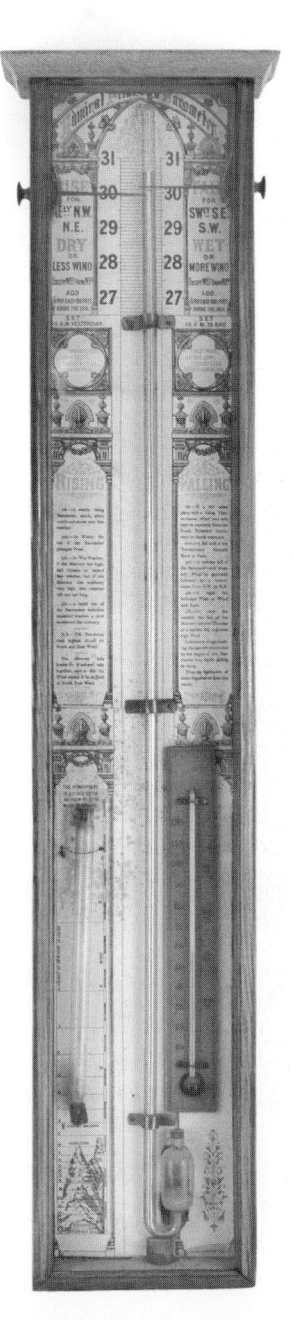

Figure 5.28 Simple late domestic FitzRoy barometer in oak case, c.1900.

Figure 5.30 Detail of top of scales of barometer illustrated in figure 5.29.

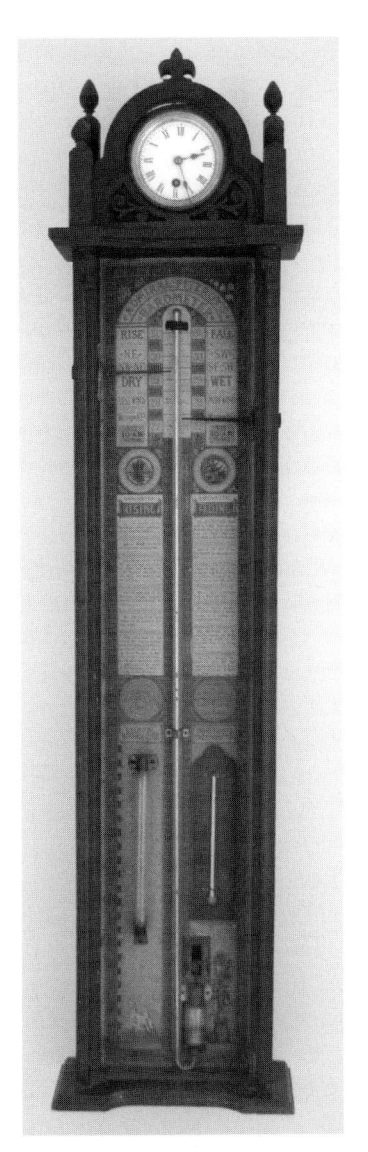

Figure 5.29 American walnut-cased domestic FitzRoy barometer with decorative charts and clock, c.1895.

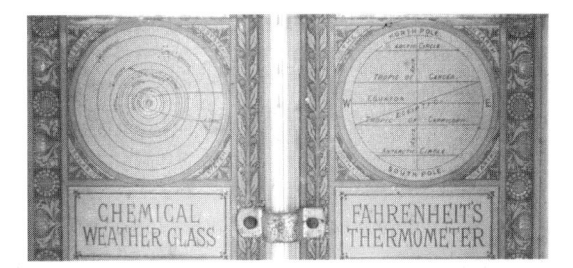

Figure 5.31 Detail of barometer in figure 5.29, showing diagrams of the solar system and the Earth.

Just when an item stops being made and then is re-made as a reproduction is difficult to determine. Figure 5.32 is a reproduction FitzRoy barometer from about the 1960s and has no doubt the same scales as I have tucked away somewhere which were used on some barometers in the early twentieth century. It is now an old barometer and looks quite authentic, even with a boxwood thermometer scale often only found on early barometers. The wood case is slightly thinner and simpler than its predecessors and the clip on the bottom of the tube is not made as the old ones were. Figure 5.33 is a newer barometer, probably from the 1970s, though the charts have faded. The makers, probably O. Comitti of London, have used the original design plugging device, although the boxwood thermometer scale has made way for an aluminium printed thermometer scale; the indicator knobs are larger and heavier than found on original barometers. Figure 5.34 shows an Oxford-framed replica FitzRoy barometer from the 1990s, and figure 5.35 is a modern, narrow-cased version which seems to appeal to people today more than the Oxford-framed style, which is regarded as looking 'very Victorian', which, of course, it is meant to!

Not often regarded as true FitzRoy barometers, but sporting FitzRoy's 'Special Remarks' on printed cards below the main dial, is a range of instruments known as Royal Polytechnic barometers, produced solely by Joseph Davis and Co. of the Royal Polytechnic Institution, Kennington Park Road, London. Perhaps the commonest of this type of barometer is illustrated in figure 5.36, although this instrument requires major restoration: it has no bezel or glass, the setting knobs are missing and the tube is broken off; even the chart, having been exposed for many years without glass, is very dark and may require replacing.

According to Edwin Banfield in *Barometers: Stick or Cistern Tube* (1985), Davis and Co. produced these barometers between 1870 and 1885. Certainly, some may have been produced later by other manufacturers, but these dates coincide with many examples I have handled. The special feature of the Royal Polytechnic barometer is the main dial: this has two indicating pointers, which, once set to the mercury level, point also to a prediction for either summer or winter, and can be quite accurate if used correctly. They frequently have twin thermometers, one Fahrenheit and one Centigrade. As in the model illustrated, the bulbs of the thermometers are often pear shaped and the scales are made in silvered engraved brass. Some Royal Polytechnic barometers have thermometer scales which are in boxwood and some have a storm glass and only one thermometer; others feature 'Admiral FitzRoy's Storm Warning Signals' (figure 5.37).

The Royal Polytechnic barometer illustrated in figure 5.38 is a large

FitzRoy's Barometer Legacy

Figure 5.32 Reproduction domestic FitzRoy barometer in oak case, c.1960s.

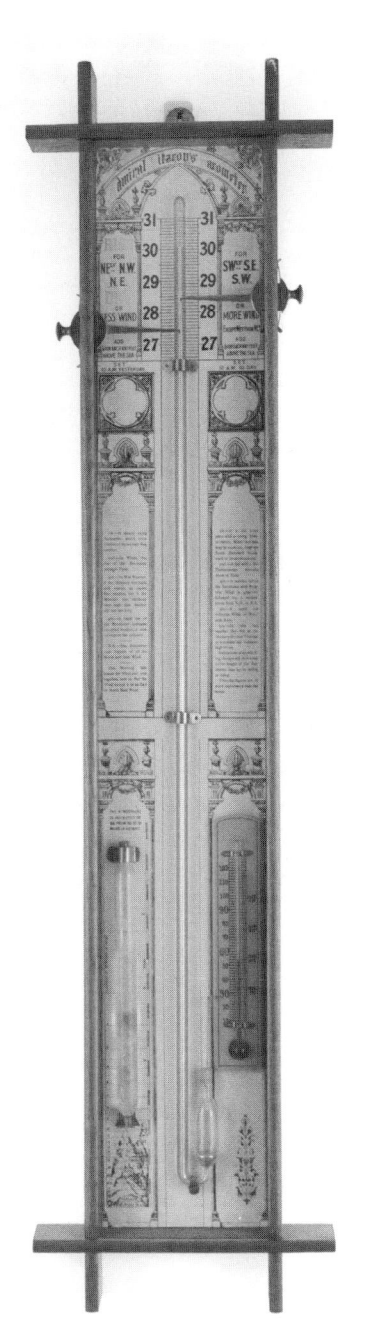

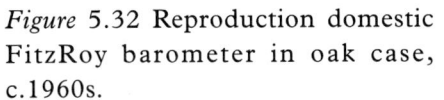

Figure 5.33 Reproduction domestic FitzRoy barometer in oak case, c.1970s, with plugging device.

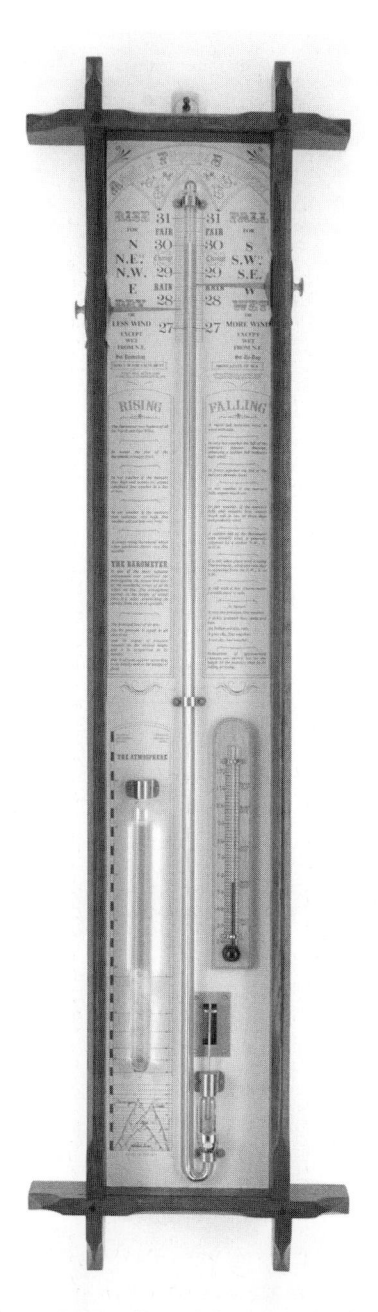

Figure 5.34 Reproduction Oxford-framed domestic FitzRoy barometer in oak case, c.1990, by Barometer World Ltd.

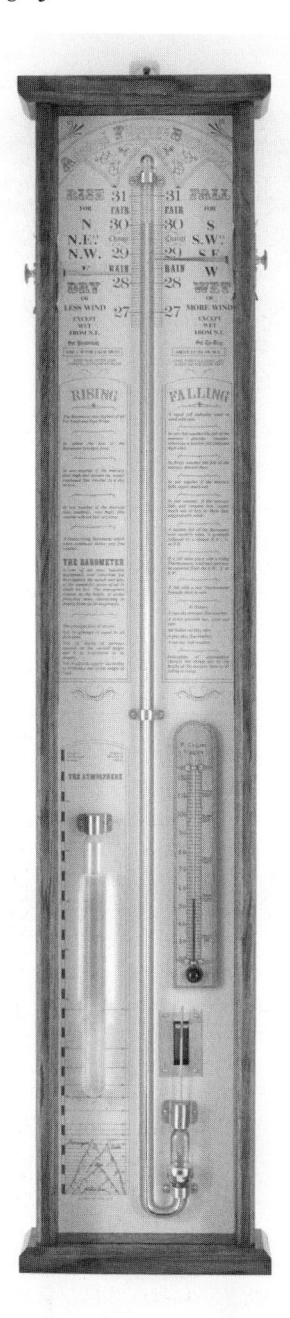

Figure 5.35 Reproduction domestic FitzRoy barometer in oak case, c.2000, by Barometer World Ltd.

Figure 5.36   Royal Polytechnic barometer in oak case by Joseph Davis and Co., c.1880.

Figure 5.37   Storm-warning signals as featured on some Royal Polytechnic barometers.

mahogany instrument in need of some repair, but with basically the same internal design. This barometer looks worse than it is: the storm glass and thermometer scale are both missing, along with the setting knobs, but the case is basically sound, with just the top finial area damaged, and the wood needing a sympathetic clean and wax polish. The main area of concern, however, to restore this instrument correctly is the very large diameter glass tube which is broken and needs replacing. Obtaining glass tubing of this diameter (18 mm) is difficult as none of the glass tube manufacturers in Europe (none left in the UK) makes anything larger than 15 mm diameter with a small bore. Sometimes a compromise may be necessary or the use of old stocks of glass if any can be found. We built a glass furnace a few years ago in an attempt to see how such large-diameter glass tube was originally made; it was a useful experiment and many of the problems became apparent. Given time and funding, we are confident that very large tube glass could indeed be produced by hand again.

Figure 5.39 shows another variation on the Royal Polytechnic barometer, the American Forecast Barometer, again by Joseph Davis and Co. but with some differences in design. The case is specially made for the design: the fretted panel is behind the glass and surrounds the silvered brass thermometer scale. The base of the case contains a horizontally mounted compass (just visible); the two indicators are supplemented by seven small, hemispherical dials (just below the indicating knobs) to record the barometer readings for a week by turning each manually with a key. This example is in perfect order and the dial, which is printed on a silvered type of card, can be seen in more detail in figure 5.40.

FitzRoy's barometer 'legacy' in terms of his weather words also made the transition from mercury to aneroid barometers. One 'find' I had in a dealer's warehouse many years ago is the aneroid barometer shown in figure 5.41, which shows the change from mercury barometers in veneered pine cases to solid wood aneroids. This model is an aneroid in a walnut veneered case by Marratt and Ellis of 63 King William Street, London Bridge, whose instruments are always of fine quality. This one probably dates from the 1860s, being one of the earlier banjo aneroids with silvered scales. It is of special interest here as it has FitzRoy's rhyming words on the dial ('Long Foretold Long Last' and 'Short Notice Soon Past'), not words that we commonly see in isolation (figure 5.42). A good Victorian, brass-cased, 8-inch dial aneroid of around 1870 is shown in figure 5.43 as an example of the thousands of early aneroids that used FitzRoy's weather indications on many different sizes and designs of dial; this one is very pleasant.

Figure 5.38 Ornate mahogany Royal Polytechnic barometer in mahogany case, c.1885.

Figure 5.39 American Forecast barometer in oak case by Joseph Davis and Co., c.1880.

Figure 5.40 Dial of the American Forecast barometer shown in figure 5.39.

Figure 5.41 Walnut aneroid barometer by Marratt and Ellis of 63 King William Street, London Bridge, c.1860s.

Figure 5.42 FitzRoy's rhyming weather words on the dial of the barometer shown in figure 5.41.

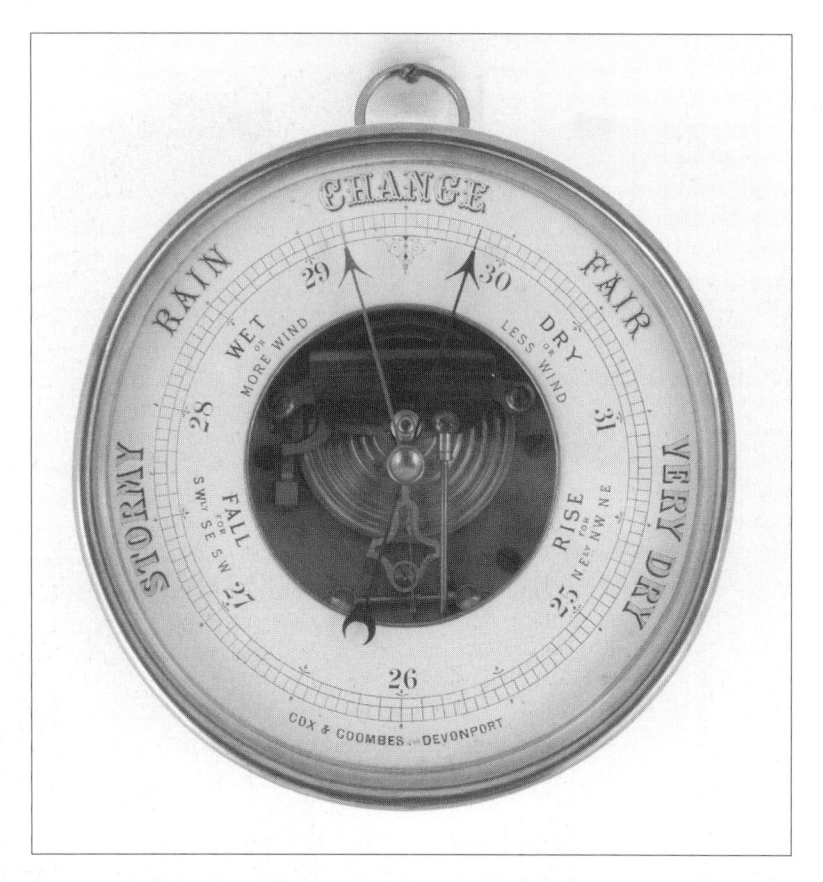

Figure 5.43   Brass-cased, 8-inch dial aneroid with FitzRoy's scale words on the dial, c.1870.

Less common are finer stick barometers which include FitzRoy's scale words. The large percentage of stick barometers with a FitzRoy connection are oak and late models of robust construction. Figure 5.44 illustrates a mahogany veneered barometer by Dollond of London, which may perhaps have been made shortly after FitzRoy's death or perhaps even before. The scale words (see figure 5.45) are not a common feature on domestic stick barometers and one can imagine that the design may have been considered rather 'vulgar' for fine pieces of furniture like this. Certainly it is a nice example; even the set key is made of tortoise shell.

So many different sorts of barometers can be found with a FitzRoy connection that it is impossible to illustrate all the examples here. A typical inlaid barometer of the Edwardian period, in rosewood veneer with white glass scales, is shown in figure 5.46. On its dial is the condensed form of

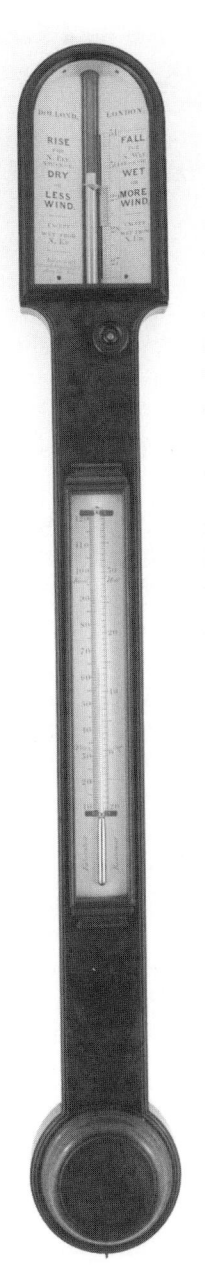

Figure 5.44 Mahogany stick barometer by Dollond of London, c.1870.

Figure 5.45 Scale of barometer in figure 5.44, showing FitzRoy's weather words.

FitzRoy's scale words so often found on late nineteenth- and early twentieth-century aneroid barometers: 'FALLS FOR WET OR MORE WIND, S.WLY S.E. S. W.' and 'RISES FOR DRY OR LESS WIND, N.ELY N.W. N. E.' (figure 5.47). The same words appear up to the Second World War, but seldom, if ever, after. An example with the same condensed form of FitzRoy's words on the dial is the brass-cased and oak-mounted ship's barometer by John Barker and Co. Ltd of Kensington shown in figure 5.48, which is typical of the late 1920s and early 1930s.

In Negretti and Zambra's *Treatise on Meteorological Instruments* (1864), there is also mention of small-sized and pocket aneroid barometers which FitzRoy appears to have requested them to make:

> The patent for the Aneroid having expired, Admiral FitzRoy urged upon Messrs. Negretti & Zambra the desirability of reducing the size at which it had hitherto been made, as well as of improving its mechanical arrangement, and compensation for temperature. They accordingly ... succeeded in reducing its dimensions to two inches in diameter, and an inch and a quarter thick.

Figure 5.49 shows the type of instrument they describe. Soon after, they reduced its size even more to what we now consider to be the 'pocket' barometer (figure 5.50), or, as they describe it: 'The smallest size can be enclosed in watch cases ... so as to be adapted to the pocket.'

FitzRoy was well aware of the usefulness of the pocket-sized aneroid. Negretti and Zambra quote him as saying that: 'Aneroids are now made more portable, so that a pilot or chief boatman may carry one in his pocket, as a railway guard carries his timekeeper.' And in the 1862 report of the Meteorological Office, FitzRoy wrote: 'Aneroids have been also improved, as suggested here, and some are now made – small enough for the pocket – useful for comparing heights in reconnoitring, besides being reliable as good weather glasses.'

I have not yet come across a pocket barometer sporting any of FitzRoy's scale words, despite seeing the larger, $4\frac{1}{2}$-inch diameter, brass-cased barometers with the simple rise and fall words as on many barometers. The simple pocket barometer, perhaps due to its size, seems not to have included FitzRoy's words, though it is possible that such an item may turn up in the future.

All the different instruments described in this chapter demonstrate how FitzRoy left his mark on barometers in one form or another. The reader will also find many other barometers that include FitzRoy's weather words or sayings – discovering barometers of different designs is part of

Figure 5.47 Dial of barometer in figure 5.46, showing FitzRoy's condensed weather words.

Figure 5.46 Inlaid rosewood aneroid barometer, c.1910.

Figure 5.48 Brass-cased and oak-mounted ship's barometer by John Barker and Co. Ltd of Kensington, c.1925.

Figure 5.49 Negretti and Zambra were the first to make these 'pocket'-type aneroid or mountain barometers.

Figure 5.50 Normal pocket-type barometer by Negretti and Zambra, c.1895, developed from small aneroids.

the joy of collecting. 'FitzRoy' barometers can claim a significant place in the history of the barometer, both for their scientific and domestic use. Their present popularity and the esteem in which they are now held by dealers and collectors alike is a fitting testimony to the important work of FitzRoy and his weather forecasting activities.

6 Buying a FitzRoy Barometer

There is much to be considered when buying a FitzRoy barometer, especially for the amateur. Perhaps a major consideration for most people will be the budget that they are willing and able to commit to such a purchase, which, of course, will rely on a number of interacting factors. First, what type of FitzRoy barometer are you looking for? Is it an original storm or sea-coast type or one of the later domestic barometers inspired by FitzRoy and sold in large numbers after his death in 1865? The FitzRoy marine barometer will be the hardest to procure as it does not sport any marks to indicate that FitzRoy had anything to do with it. It is the design alone that puts this barometer into contention. Lists of the many marine barometers – each numbered – that were issued during FitzRoy's days at the Meteorological Office survive in the National Archives, and so it may indeed be possible for the purist collector of naval instruments to acquire one with a little patience and hard work.

FitzRoy Marine or 'Gun' Barometers

Since the availability of FitzRoy marine barometers is very limited, condition may have to be sacrificed more than is necessary with the plentiful domestic FitzRoy varieties still available (see below). That is not to say that the would-be collector should abandon prudence and 'buy regardless' the first marine instrument that he or she comes across. Instead, give a bit more consideration to the feasibility of having the barometer restored, if needed, rather than perhaps expecting to be able to buy one in such good order that it only needs hanging on the wall.

Marine barometers are invariably made of brass and, with rare exception, were originally painted black with polished and lacquered brass mounts, including the glass scale cover supports, the vernier knob, the screws for the gimbal mounting, the thermometer loops, the hanging land hook and various screws, the gimbal ring and main trunk being matt black finished. On earlier models, perhaps before the 1920s, the reservoirs were more commonly made from cast iron and were black painted; with age, they tend to show outward signs of rust. Around the 1930s, these reservoirs were replaced with stainless steel, which, of course, survives nearly as originally made. It is often the opposite case for the glass cover and mercury tube, which, after being returned to stores for another repair, may have been deemed to be in too bad a condition to repair – or perhaps too old – and so were superseded by another, more modern replacement.

Many of these working instruments were sent back for recalibration, and when this was done, the year of calibration was engraved on the instrument. An instrument may have been recalibrated as often as five or six times, as can be seen in figure 6.1 which shows the engraving of Meteorological Office instrument number 3473, first calibrated in 1941 and recalibrated in 1945, 1952, 1956, 1963 and, finally, 1971. This is, of course, a very easy way of dating such instruments. Early ones were not dated like this, although their certificates were. If the barometer you are buying comes with its original carrying case, it will have the original certificate of testing by the National Physical Laboratory (NPL) pasted into the lid, as seen in figure 6.2. Note that the instrument number should match the number on the barometer. The chances of buying an original barometer issued by the Meteorological Office during FitzRoy's time as chief (1854–65) is not great, but over years of collecting I have come across a number of these items and there must be more out there which may come onto the market in the future.

Probably the main consideration with such instruments is whether they are capable of being sensibly restored, if needed. The glassware can normally be replaced, although this will cost several hundred pounds if done professionally. The brass may well have corroded after the lacquer has deteriorated but can – unless exceptionally corroded – be cleaned and re-lacquered. The brass trunk and fitments, originally black, can be re-painted, although, if there are only reasonable signs of wear and tear, they would be better left as signs of age and use. Sadly, among the many people who have handled these items since they came onto the open market there will have been some who decided that these instruments look better polished back to brass. It seems that many dealers prefer this look as it makes the barometers appear 'more nautical' and thus, perhaps, more saleable, regard for originality giving way to a desire for profit.

The scales and vernier can often be a critical part of the instrument, and accurate division by the high quality of the engraving is a key factor in these instruments passing the NPL test in the first place. A careful look at the scales may show greying areas or spots which are indicative of mercury contamination corroding or discolouring the silvering on the brass. This can often be removed if only slight, but heavy corrosion of a number of years may be difficult to remove completely – and caution should be exercised before purchasing such an instrument.

Almost worse than a mercury-corroded scale is a polished scale, again often caused by the unscrupulous dealer's zest for polishing everything

Buying a FitzRoy Barometer

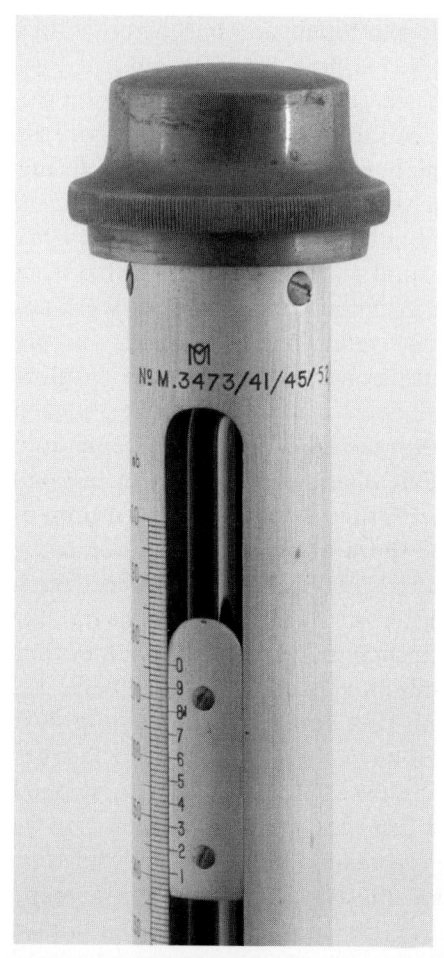

Figure 6.1 Engraving on Kew marine barometer number 3473 showing recalibration dates.

Figure 6.2 NPL certificate of instrument 3473 (shown in figure 6.1) inside lid of carrying case.

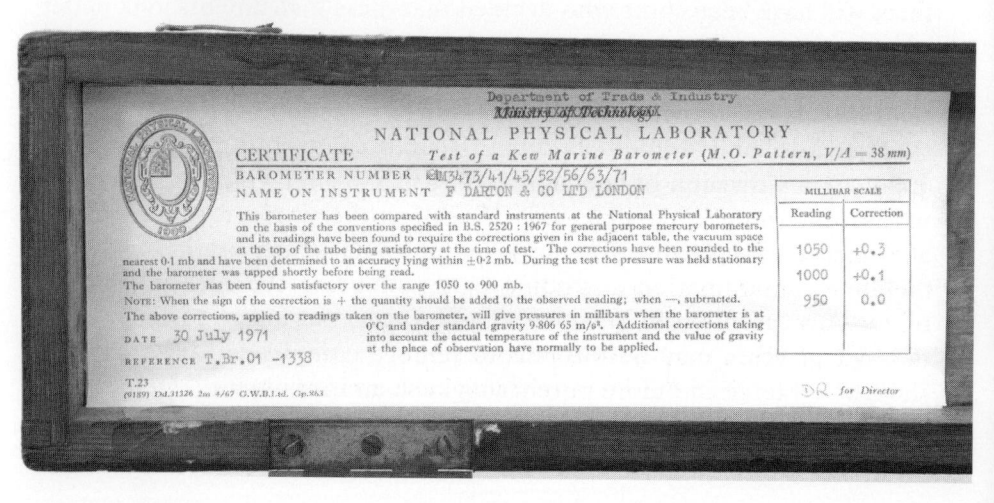

brass to a gleaming 'nautical' yellow or sometimes the result of the unknowing owner cleaning the grubby scales until brass shows through. Like frequently polished warming pans, the engraving becomes rounded and, in the worst cases, begins to disappear. The only remedy is to re-engrave the scales after thorough cleaning of the brass, but this is something that is unlikely to be financially viable and to be avoided. One of the worst barometers of this type that I have seen was brought into our workshop some years ago, having been found whilst diving. It had barnacles growing on the metal case, the brass was green with verdigris, the lower reservoir section had been broken off and the whole tube had a considerable bend in it. On the grounds of cost, it was not worth consideration as a restoration project.

Another factor that should be taken seriously is the number of fakes on the market. I have personally seen several of these items which, on first glance, look 'right', but on closer examination turn out to be poorly made copies. These may originate from India: one item we were asked to restore was bought in India by an airline pilot some years before. It seems that an original item has been copied, but the very knowledge of what the instrument is and how it works has been lost. A few I have seen have had brass reservoirs instead of iron or stainless steel (which are both unaffected by mercury), and all the ones I have come across have not had mercury columns inside and were probably not made with them; instead, there may sometimes be a piece of glass or an empty tube which could not work and is not fitted correctly. Another example was a copy of an early design by Newman with a revolving cistern for transport and restrictions for slowing the movement of mercury but no possible way of inserting a mercury tube. Often these items will have the name of the maker copied from the original instrument.

The first thing to look for when checking the authenticity of a marine barometer is the engraving on the scale. The Indian copies frequently have very poor engraving with lines that are not straight or equal in division or depth; the letters and numbers have been stamped on and are not aligned correctly. Once aware of this, it is often easy to spot. I should not want the reader to think that the market is awash with such fakes, but to be on their guard against this possibility.

Of the early 'gun marine' barometers as illustrated by Negretti and Zambra (see figure 3.3), and of which I have only seen a couple of possible originals, there are many modern copies in the same style as that illustrated in figure 6.3. These are frequently sold as original barometers and no doubt there are many happy owners unaware of the truth. Anyone with an eye for the quality of the original instruments can soon see that the poor

Buying a FitzRoy Barometer

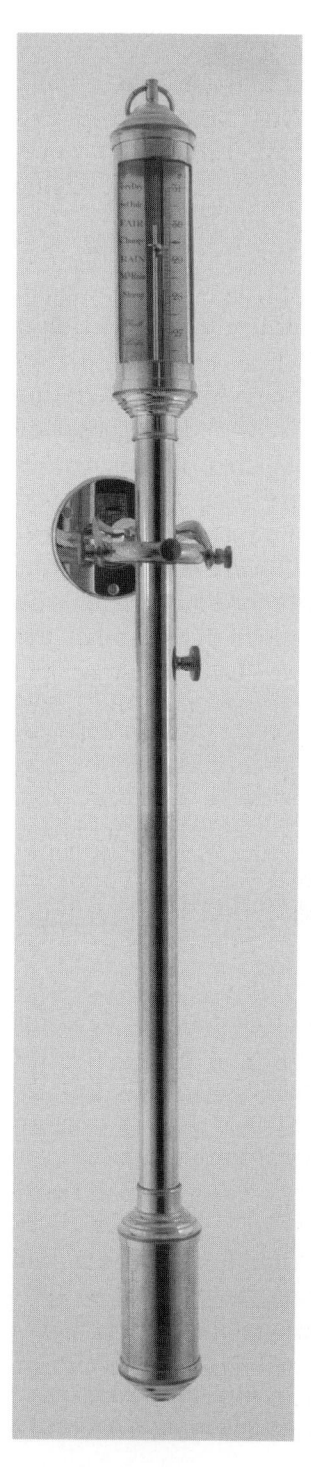

quality of the brass work, the poorly fitting vernier, the badly made hanging ring, the poor quality engraving (actually a chemically etched process) and the often plastic 'glass' cover are no match for the real marine instruments. On looking inside, there is seldom any resemblance to the original design but a modern arrangement of glass tube and sometimes a small wooden reservoir.

The names can vary, but the ones that turn up most commonly are engraved with a number and 'Desterro, Lisbon', whom I believe is the maker. I do not think for a moment that these instruments are made to deceive but are made as modern, brass-cased barometers. Some of these models have been around for over 30 years, and the firm is to my knowledge still in business. A year ago I had one such barometer come in for repair from a carrier who had broken the tube in transit. It appears that the owner had paid a considerable sum of money for it as an antique – well, buyer beware! There are a number of other modern barometers on the market, so take care when considering both price and quality as these two factors can indicate that these instruments are not original pieces.

FitzRoy Storm Barometers

Storm or sea-coast barometers are perhaps the main interest of mine and many would-be collector or barometer owner. They show a direct link with Admiral FitzRoy's request to Negretti and Zambra (see chapter 4), and even if an original one issued during FitzRoy's time at the Meteorological Office is unlikely to become available, there are a number of the same design that were issued or used after his death that he would have recognised, as well as many that were

Figure 6.3 Modern brass marine barometer by 'Desterro, Lisbon'.

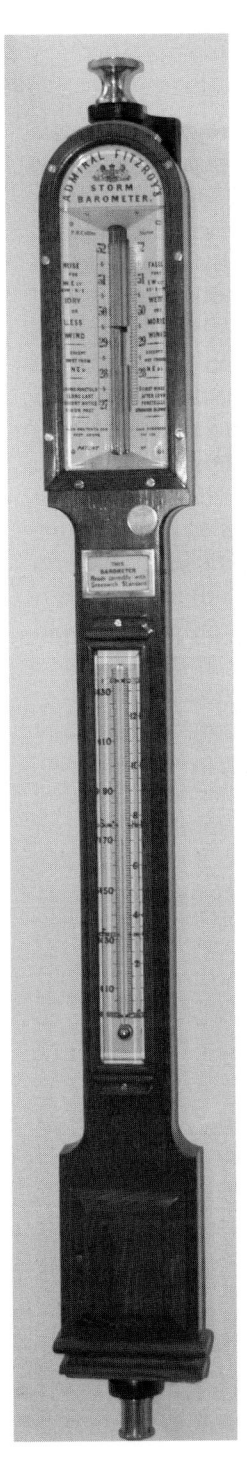

sold for domestic and institutional use that come onto the market. Having worked on a number of these barometers on behalf of the RNLI, which still maintains many for historic reasons around the coast, I have found that many would have had large hanging brackets (see figure 6.4) to enable the barometer to be secure and stable and to be turned, if required, to allow reading from different angles or from inside or outside if displayed in the window of a building. Many others were also used with the normal design of hanging plate: perhaps cost was a factor in this, as there were many bought by public subscription and donations, as well as by the Board of Trade.

The normal method of numbering the instrument by the makers Negretti and Zambra was on the lower section of the right-hand porcelain scale. The number would have been individually hand-lettered, and will frequently have disappeared over the years as this method of numbering was seldom as permanent as the lettering of the scales. Later ones have Meteorological Office instrument numbers for both the barometer and the attached thermometer; normally this number is on the scale of the barometer and there is a separate number on the thermometer. We know where many were placed around the coasts between FitzRoy's day and the Second World War. Many, I am told by 'old boys', were taken down during the war in anticipation of bombing and many were never returned after the war, but languished in a loft or store or were acquired by local residents as these instruments were regarded as old and out of date and were no longer wanted.

There is a surviving list of barometers used to supply readings to the Meteorological Office which shows the location of 105 instruments in Devon and Cornwall alone. They were possibly not all sea-coast barometers, but certainly the majority of them were. Only a few still exist in their original positions: for

Figure 6.4 Modern storm barometer mounted in heavy cast brackets.

obvious reasons, I will not highlight the location of individual instruments. Figures 4.4 and 4.5 show how these instruments were displayed in small fishing villages (see chapter 4). Many were housed in simple glazed boxes; often the box was built into a wall.

An obvious feature to look for when acquiring a storm barometer is the weathering of the wood. If the barometer has been used near the coast, even if behind glass, then the wood, after many years of variable weather conditions, moisture and temperature, will shows signs of ageing. In examples that have been exposed for a very long time, the wood will appear rough. This is part of the natural ageing of the wood: the hard grain tends to remain, while the soft wood between gradually wears away, producing what could be, in some instances, a very appealing condition for a storm barometer, though not usually admired in good domestic pieces of furniture. In general, storm barometers available for purchase will be in a better condition than this, many of them having been housed indoors, and the polish, whilst aged, will still maintain some colour and surface finish. As with all antiques, the finish of the polished wood is important, and a good colour and patina are to be desired as a general rule. Colour can vary considerably as a number of different colours were possible and finishes varied over the years. I have seen storm barometers ranging in colour from a very dark oak to a very bleached light oak. Often golden oak can look the most attractive, but individual preference, of course, will vary.

The condition of the wooden case is very important. I have seen many barometers that have been so poorly repaired in the past that their value has been significantly reduced. Check to see that the various mouldings and wooden parts look original and that no odd pieces of wood appear to have been substituted. Without exception, traditional storm barometers were made of oak for public display. Not until later, when sold by Negretti and Zambra for private use, were a variety of cases made using mahogany or other wood, as the customer may have requested, and scales of ivory or silvered brass were used instead of porcelain as specified by FitzRoy.

As time went on, many other manufacturers started making barometers of a coastal pattern. These varied considerably in style and quality. Whilst good-quality manufacturers could and often did produce first-class instruments, there are many examples of, if not poor quality, then certainly not as high a quality as most Negretti and Zambra instruments. I say *most* Negretti and Zambra instruments because, unlike their domestic pieces, I have seen a number of their storm barometers that are surprisingly rough in comparison with their other instruments. I suspect that, being a rugged instrument, less attention was often paid to the case of a storm barometer than to the mercury tube and scales: these instruments were designed as utilitarian instruments, not decorative pieces. There was no need to make them attractive and so, whilst the design did not vary, you can see that some of the oak cases are simpler and slightly cruder when comparing the shape and curves of the wood. Perhaps I am extra-critical, but years of handling an object gives one an appreciation for its finer details.

I suspect that, as fine detail was not important, the cases were made by craftsmen of more robust skills who were more akin to carpenters than fine cabinet makers. This can frequently be evidenced in different examples and specifically when comparing the thermometer boxes. The thermometer scales are made of porcelain and, as we have discovered by experience, porcelain scales are not very flexible. Once made, you have to make the cases to suit, and any variation in the case cannot simply be accommodated by engraving the scale so that the numbers are in the correct position, as might be possible with individually hand-engraved scales. When the scales are made in quantity and the wooden cases do not match, the thermometer scales have to be broken off at the top and this will often be where the coat of arms and name 'Negretti & Zambra' are situated. To get over this obvious mis-fitting, you will find that some thermometers have brass plates fitted at the top and bottom of their scales (see figure 6.5 for the top of the scale) to 'cover up' this rather crude correction. It is not a later repair or fault, but a manufacturing technique commonly found on this style of barometer.

Another feature to check is the vernier to see if it is working correctly. This is a small sliding device alongside the scale that enables the barometer to be read to within a hundredth of an inch; some later ones have two verniers – for today and tomorrow – but this is not generally the case on storm barometers. The vernier is operated by a small brass key that locates into a hole with a square pin inside to enable the

Figure 6.5 Thermometer showing brass plate covering broken top of porcelain scale.

'rack', a toothed gear, to move up and down when the key is turned. Provided there is a key in place – many get lost over the years – this should turn and the vernier plate move reasonably easily up and down; there should be no hard points or grinding noises, and at no time should the vernier, if left unheld by the key, drop on its own. Regrettably, the design of these verniers is not as good as could be designed today; the wood case shrinks, even when seasoned wood is used, and it will be found on many that some adjustment of the vernier is necessary. At worst, the racks will have some teeth stripped off, and a replacement rack or a careful repair will be needed to ensure good operation. Please do not attempt to move the vernier on a barometer in an antique shop or exhibition without asking the owner first. He or she may prefer to show you its operation and may already know whether it is working or not, which will no doubt be reflected in the price.

Broken scales, or scales that have been poorly glued after being broken, are a serious factor, not to be lightly dismissed when considering a purchase. They can be restored well if a skilled restorer is used, but beware of poor-quality restoration as this can spoil a potentially good barometer. A straight crack in the porcelain scale can more easily be repaired if it has not been glued with super glue. Cracks in many places, which require the divisions and words to be re-lettered, will not only cost more to restore but be a potential problem in the future.

Another important factor is the mercury tube. If it is obviously broken or a wrong replacement (for example, visibly too small) then a replacement tube will be necessary. Most of the tubes in these old barometers have diameters of approximately 18 mm and that size of glass is not available as it was a hundred years ago. Many dealers use smaller tubes or other methods of fixing the problem. Discuss this before buying as it could easily affect the value of the barometer and how much you should pay for it.

If the tube of mercury is apparently unbroken, seldom will it be in good order if of any real age. The leather diaphragm may be so old that it will be sensible to replace it; it may be leaking slightly, and even if 'working', a close examination of the visible part of the tube should be undertaken. Most old tubes will have taken on a grey, murky appearance, especially where the mercury rises and falls regularly; if excessive, it may not be possible to clean easily or at all. The old tube will, in a majority of cases, need dismantling, cleaning and refilling with triple-distilled mercury. The old practice of filtering the mercury is a waste of time and effort as filtering only removes the particles of dross and not the dissolved elements that are often absorbed into the mercury from the glass and air; if the mercury

Figure 6.6   Dirty barometer tube.

was not pure when filled, there could be extra dissolved pollutants in it. The mercury may look clean but it should be chemically pure to give many years of trouble-free use and enjoyment. Figure 6.6 shows part of a dirty tube before cleaning.

On close examination of some tubes just above the top of the boxwood cistern, you may notice a grey as well as a sparkling appearance. The sparkling, shiny spots are actually small bubbles of air which gradually build up and move up to the top of the tube drawn by the vacuum and by gravity. It may be possible to tip the barometer tube up and allow the bubbles of air that have accumulated in the top of the tube to run up and into the cistern area. However, do not be misled by this because, although it may seem a simple solution, these air bubbles will frequently reappear. They enter into the tube because of the build-up of dirt around the open submerged end inside the cistern. This small build-up of dirt actually creates an entrance for very small particles of air to be drawn slowly up into the tube by air pressure, and as air moves more easily than mercury, these very small air bubbles gradually increase until, with a considerable build-up of dirt, you find a tube as illustrated in figure 6.6 where large bubbles of air are squeezing between the glass and the mercury aided by the coating of contamination.

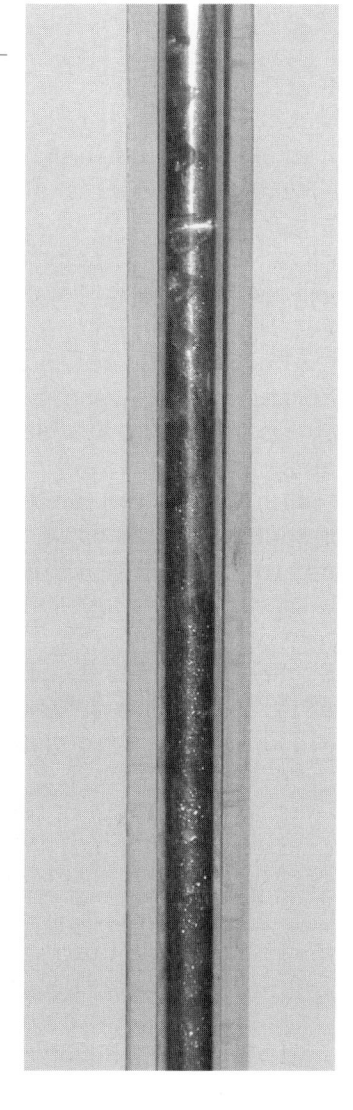

In this example, there is no alternative to serious cleaning of the tube or possibly replacement. It is often the fact that old tubes are effectively etched by years of contact with mercury. Presumably some of the chemical composition of the glass is absorbed by the mercury which adds to the contamination. Even when cleaned, the tubes remain microscopically etched which allows the entry of tiny air bubbles and then the only answer is a new tube. Experience of restoring barometers like this has generally given me an 'eye' to determine whether it is worth trying to clean the tube or not, although occasionally I still get caught out trying to make do with a tube that looks clean by eye and save the client some expense, only to find that air enters in easily and the tube has to be replaced. So be warned and do not be misled by the old sales patter – 'it's just a sign of age' or 'it's working well' – when in fact there is often no guarantee or comeback and possibly an expensive replacement tube is necessary to have a fully working barometer.

As with the mercury tube, if the mercury thermometer is broken, this can also at present be replaced – not cheaply, but satisfactorily. However, how long this will continue is uncertain due to the decline in this type of work over recent decades as a result of the modern preference for electronic devices in industry and increasingly in the home. In addition, when it comes into force, European Union legislation to ban the use of mercury in new barometers will further erode the skills base needed to restore old barometers. Whilst there may always be skilled people able to restore barometers, the raw materials required to undertake repairs has dwindled to such a degree that it is often impossible to purchase the type and size of glass required to restore some types of barometer.

Domestic FitzRoy Barometers

As we saw in chapter 5, there have been so many different types of domestic FitzRoy barometer produced over the years that it is impossible to illustrate them all here. Generally, they will all have paper scales behind a full-length glass front: these scales may cover only the top half of the barometer or extend down its full length. And then there are always exceptions, such as the engraved, silvered, brass plates found on a few good-quality ones, or the porcelain dials and scales found on others. By far the most will be in oak or ash, but others may be in American walnut (popular in the early 1900s), as well as mahogany, which can be found not too infrequently.

I have always stated that with this type of barometer the most important single factor when considering a purchase is the condition of the paper scales. If these are damaged even partly, especially if parts of the text are missing, then the value is dramatically reduced. Figure 6.7 shows a barometer with wide sides (which incorporate indicators that are adjusted by a small key inserted into the holes) and an original thermometer and scale, with the remains of a paper scale in very poor condition. It would be impossible to restore or repair, so a replacement is the only remedy. Figure 6.8 shows a common design of barometer in an Oxford-framed case, reminiscent of Victorian framed biblical texts and pictures. The unusual pointed top to the thermometer scale is original and the storm glass is intact but its contents have evaporated. Some thoughtless 'restorer'

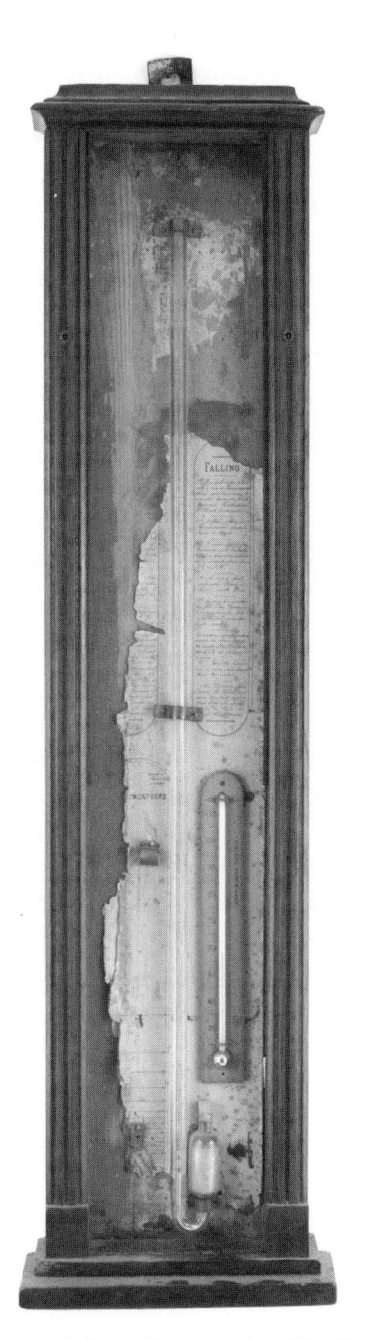

Figure 6.7 Domestic FitzRoy barometer with badly damaged paper scales, c.1870.

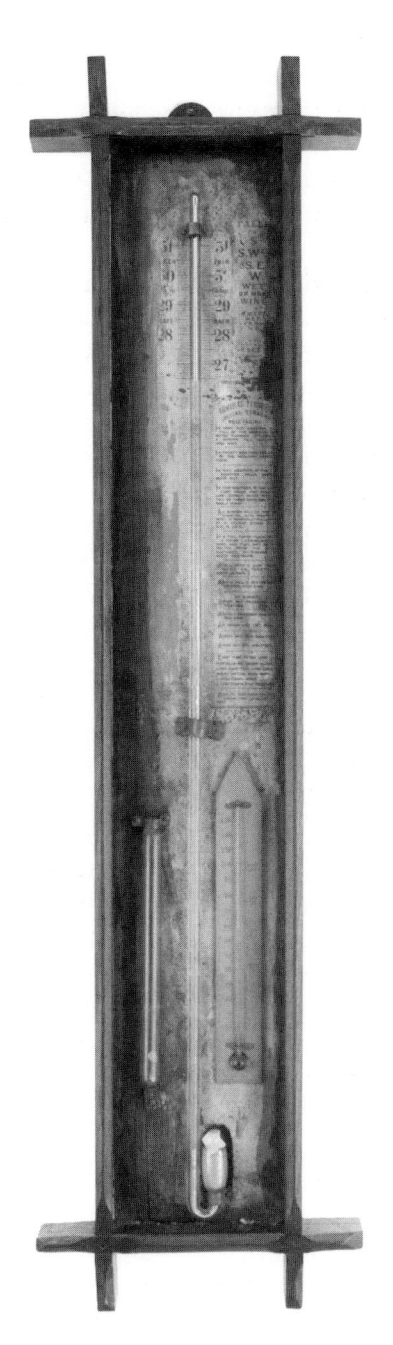

Figure 6.8 Domestic FitzRoy barometer with badly stained scales, c.1890.

has stripped the woodwork and carelessly stained it with the consequence that stain has spread over the paper scales and ruined any original 'age' that might have been retained; a replacement scale is all that can be suggested. The tube is a later poor fit as well, as can be seen by the gap around the bulb.

Figure 6.9 shows the same type of barometer with what appears to be a much better scale, but closer inspection (figure 6.10) shows heavy staining from a leaking storm glass and missing edges to the scale on the left-hand side, probably caused by silverfish. It is probably not advisable to buy a barometer in this condition unless good replacement scales are used, thus affecting the value. A number of different replacement scales can be obtained for those occasions when the owner's only choice is a replacement. However, always try to buy a barometer with good scales if you can; the other features of the barometer are more likely to be repairable. If your budget is small, of course, a poor chart may lower the value enough to tempt you and you will still end up with a worthwhile item, though less 'original'.

As with storm barometers, the condition of the wooden case is also important, though you may find that it has been subject to poor repairs over the years. In the days when the collector's preferred choice was a fine Georgian stick or wheel barometer, the FitzRoy domestic barometer was seen as a very poor relation. Most restorers at this time would not have given them a second glance; they were just not worth much. So the owner, either from choice or necessity, may have had to effect repairs in an amateur way as if repairing any piece of low-value, second-hand furniture. The result is a legacy of poorly restored cases, often with unfinished replacement parts or pine used in place of oak or mahogany. If you think what these items are worth today and reflect on their value, say, in the 1960s, then it is more understandable, for they were seldom if ever collected then or bought by antique dealers.

On nineteenth-century country furniture, this sort of repair may be seen as the work of local carpenters and add, if not to the value, then at least to the interest of the item. With barometers, the market is not so forgiving, and collectors will generally, and I believe rightly, expect a reasonable quality of repair to reflect the original quality of the instrument. 'Instrument' is perhaps too 'scientific' a word to use for some of these mass-produced items, but they do have a primitive charm of their own and some examples can be very pleasing. Check the back of the barometer to see what the case looks like. They can frequently be made of the cheapest wood and show signs of warping, woodworm and cracks that can affect the paper scales. The worst cases of woodworm will greatly affect the

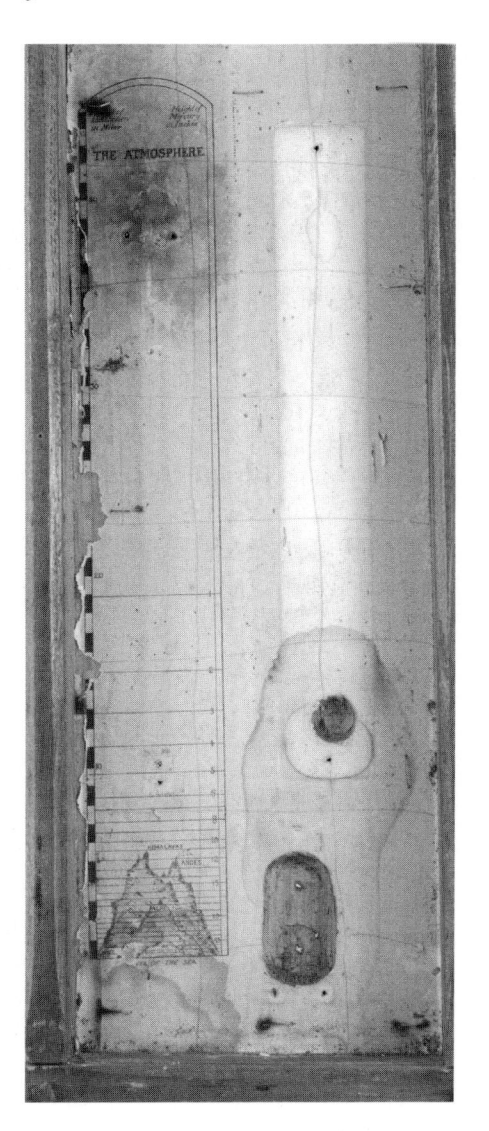

Figure 6.10 Detail of paper damage to barometer shown in figure 6.9.

Figure 6.9 Domestic FitzRoy barometer with case and scales only, c.1890.

viability of such a barometer, causing major structural replacement to be necessary.

Providing the chart and the wooden case are in reasonably good condition, we can turn next to the 'insides' of the barometer. Many will have a storm glass or camphor bottle (see chapter 5) which will have dried up over the years. This can be re-filled and re-corked by a good restorer. I would advise against attempting to make the liquid yourself. Although Negretti and Zambra's recipe can be found in my book, *Bizarre Barometers* (Collins 2004: 19), by the time you have bought the chemicals and experimented to get the right composition for a 'good' mixture, you will have spent far more time and money than you ever thought possible.

Not all of these domestic barometers had a storm glass, and I would advise against fitting one if the barometer did not originally have one. I think all of these types were originally fitted with a thermometer, mostly on a boxwood backing and frequently with a spirit thermometer tube. If missing entirely, it can be difficult to get a really good replacement, but again this should be reflected in the price and you may consider it worth the purchase on balance. The main mercury tube is important and can vary considerably. Unlike a storm barometer, the whole tube is on display, and the clarity of the mercury and the condition of the glass on the inside of the tube are often noticeable.

The tubes of old barometers will invariably be so dirty as to need specialist cleaning or replacement tubes. Take note to look carefully where the bulb of the tube fits. On so many of the barometers we have handled, someone has previously replaced the tube with a different style or shape and has gouged pieces of wood out of the case and left a pretty sore sight, all for the want of a properly fitting tube. Figure 6.11 shows a barometer that has had the wrong shape of tube fitted: instead of sourcing a specially made tube, a standard modern tube has been fitted with the resulting disappointing appearance – anyone can tell it is wrong! Although the case has not been carved out to fit the new tube, unsightly clip holes have been made when fitting the clip in the wrong position; the original holes can be seen to the right. The problem can sometimes be remedied by making a larger bulb than the original and cutting a new shape to fit and cover the damage. Figure 6.12 shows another poorly fitted tube, but no additional damage has been done and here it may simply have been an effort to adjust the reading of the barometer by lifting the tube. The charts are later replacements. Some dealers continue to bodge repairs to save money but they only waste their time and spoil these barometers for future generations.

Critical in purchasing one of these barometers will be the diameter of

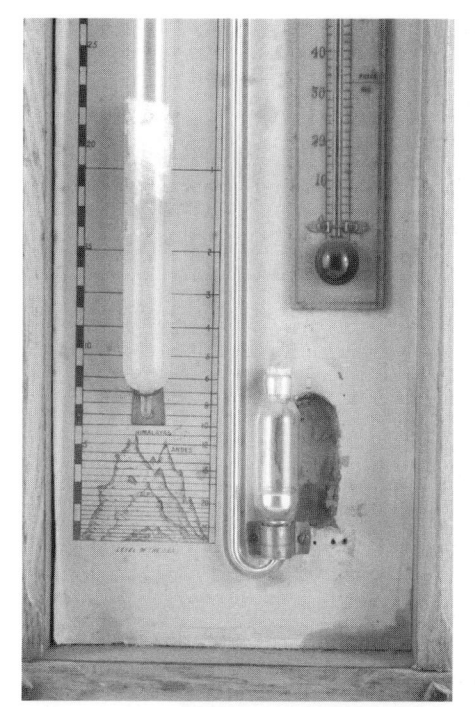

Figure 6.11 Unsightly cut-out shows where original tube was fitted.

Figure 6.12 Badly fitting tube in domestic FitzRoy barometer.

the glass tube. Many glass tubes will be 12 mm in diameter or less which is still currently available in capillary tubing. We have stocked specially made capillary tubes up to 14.5 mm in diameter, but when it comes to 16 mm and larger, which can be found on large Victorian models such as that shown in figure 6.13, then the glass is not commercially available. Despite some success in cleaning and annealing old tubes, it is a risky business and breakages can be a natural part of this work. Joining old bits of glass is a job only for a highly skilled glassblower. In most cases, if the tube is 12 mm in diameter or less it would be normal practice to replace the tube and fill with clean mercury for many years of satisfactory service.

The brass fitments, such as the sliding level indicators (see figure 6.14) or the rack-and-pinion operated type (figure 6.15), can often be damaged. Although the job is sometimes fiddly, they can normally be repaired or replaced. The rest of the brass items are normally cleaned and lacquered as is traditional on these and other scientific instruments so as to preserve them for some years. If there is significant verdigris or corrosion, some repair or replacement may be needed. The hanging plates, often made from thin tin, can be bent with misuse. If the barometer is fitted with a

Figure 6.13 Royal Polytechnic barometer, which should have a large (18 mm) tube, fitted with too small a tube.

patent transport device, this is often not working, but can with some basic skill be re-fitted.

Fisherman's Aneroids

The cases of early examples of fisherman's aneroids may have been made of bronze, but they are more likely to be a form of pewter which melts easily if you try to re-solder the hanging plates – so be warned! If dented, they can be impossible to get back into shape. The brass bezel, which is sometimes finished in black on later models, can suffer from metal fatigue and show the characteristic radial cracks around the edges; if pronounced, a new bezel may be needed which would need to be specially made – at considerable cost.

The quality of these aneroids is invariably very good; the movements are temperature compensated and accurate. The bezels, which are cast on early models, spun on later ones, are usually bolted through the case with

Figure 6.14 Manually operated level indicators in domestic FitzRoy barometer. Note the badly fitted replacement charts and poorly made piece of wood at the top.

Figure 6.15 Rack-and-pinion-operated level indicators on a domestic FitzRoy barometer.

round brass nuts to secure them in place. Often the barometers survive in good order internally but, of course, can suffer from the usual faults of aneroids. Some of these faults can be the normal result of age and corrosion, requiring a complete clean and overhaul which might require a replacement fusee chain, replacement hair spring and even a replacement capsule. Further information on the restoration of aneroids can be found in my book, *Aneroid Barometers and their Restoration* (Collins 1998). The condition of the enamel dial is important to the value of the barometer and, if damaged, can drastically reduce its worth. Repairs to enamel dials can be effected, but at a price.

For all types of FitzRoy barometer, as with any worthwhile antique, one should consider buying from a reputable dealer who has a reputation to lose and cares about the items he or she is selling. Look for the skills that the dealer has and compare the barometer you are interested in with other items on show. Is the dealer truly knowledgeable or a general antique dealer who may not have specialist knowledge about the item? That is not to say that you cannot buy a good barometer anywhere, but make sure you know what you are buying or buy from someone in whom you can have confidence and trust.

Buying in auction has become more common these days but care must be taken as once a bid has been accepted you are obliged to buy – even if you mistakenly bid for something else! There is no comeback if the item is faulty or broken, even if it breaks after the hammer falls, and, unless grossly mis-described in the catalogue, there is no chance of returning for a refund.

So, armed with a little caution and some knowledge, readers who want to buy a FitzRoy barometer should be able to find something that they will enjoy for years to come – or perhaps even start collecting them.

Bibliography

Banfield, Edwin (1985) *Barometers: Aneroid and Barographs*. Trowbridge: Baros Books.

— (1985) *Barometers: Stick or Cistern Tube*. Trowbridge: Baros Books.

Burton, Jim (1986) FitzRoy and the early history of the Meteorological Office. *British Journal for the History of Science*, 19: 147–76.

Collins, Philip R. (1998) *Aneroid Barometers and their Restoration*. Trowbridge: Baros Books.

— (2004) *Bizarre Barometers and Other Unusual Weather Forecasters*. Trowbridge: Baros Books.

FitzRoy, Robert (1839) *Narrative of the Surveying Voyages of His Majesty's Ships Adventure and Beagle*. London: Henry Colburn.

— (1860) *Barometer Manual*, 3rd edn. Issued by the Board of Trade. London: Eyre and Spottiswoode (reprinted by Barometer World and Museum, Merton, Devon, 2000).

— (1863) *The Weather Book*, 2nd edn. London: Longman, Roberts and Green.

Gribbin, John and Gribbin, Mary (2003) *FitzRoy: The Remarkable Story of Darwin's Captain and the Invention of the Weather Forecast*. London: Review.

Halford, Pauline (2004) *Storm Warning: The Origins of the Weather Forecast*. Stroud, Gloucestershire: Sutton Publishing.

Keynes, R. D. (ed.) (1979) *The Beagle Record*. Cambridge: Cambridge University Press.

McConnell, Anita (2006) Will the true originator of the storm glass please own up. *Ambix*, 53 (1): 67–75.

Marquardt, Karl Heinz (1997) *Anatomy of the Ship HMS Beagle: Survey Ship Extraordinary*. London: Conway Maritime Press.

Mellersh, H. E. L. (1968) *FitzRoy of the Beagle*. London: Rupert Hart-Davis.

Middleton, W. E. Knowles (1964) *The History of the Barometer*. Baltimore, MD: The Johns Hopkins University Press (reprinted Trowbridge: Baros Books, 1994).

Moon, Paul (2000) *FitzRoy: Governor in Crisis 1843–1845*. Auckland, New Zealand: David Ling Publishing.

Moorehead, Alan (1969) *Darwin and the Beagle*. London: Hamish Hamilton (reprinted by Book Club Associates 1978).
Negretti & Zambra (1864) *A Treatise on Meteorological Instruments*. London: Negretti & Zambra (reprinted by Baros Books, Trowbridge 1995).
Nichols, Peter (2003) *Evolution's Captain: The Dark Fate of the Man who Sailed Charles Darwin around the World*. New York: HarperCollins.

Index

Adie, Alexander, 11, 12
Adie, Patrick, 41, 42, 43, 48, 49, 50
'Admiral FitzRoy's Barometer', *see* domestic FitzRoy barometer
'Admiral FitzRoy's Remarks', 85, 108; *see also* scale words
American Forecast Barometer, 112, 113
Babington, Thomas Henry, 67, 91
Barker and Co. Ltd, John, 118, 120
Barometer Manual (FitzRoy), 55, 56, 58–9, 62, 84
Beagle, 2, 4, 5–7, 9–11, 15–16, 18, 37, 39
Beaufort, Sir Francis, 6, 15, 40–1, 42; wind scale, 6–8
Beechey, Captain Frederick William, 19
Board of Trade, 20, 25, 28, 50, 55, 57, 66, 127
British Association for the Advancement of Science, 18, 23
Burgoyne, John Fox, 19
calibration, 123, 124
camphor glass, *see* storm glass
Casella, 43, 51, 53
coastal barometer, *see* storm barometer
Comitti, O., and Son Ltd, 79, 108
Darton, F., and Co., 50, 51
Darwin, Charles, 7, 9
Davis, Joseph, and Co., 100, 101, 108, 111, 112
Dollond, 74, 78, 79, 80, 83, 116

domestic FitzRoy barometer, 84–8, 92, 94–108, 132–8
engraving, 123, 125
fakes, 125
fisheries barometer, *see* storm barometer
fisherman's aneroid barometer, 74, 76, 78–83, 138–40
FitzRoy, Maria, 17, 39
FitzRoy, Robert, 38; *Beagle* voyages, 2, 5–16, 37; death, 39; early life, 1–2; family, 17; governor of New Zealand, 16; at Meteorological Office, 19–21, 35–7; and RNLI, 62
FitzRoy marine barometer, 46–52, 122–6
FitzRoy's weather words, 74, 84, 112, 115; *see also* scale words
forecast, 29–32
Fuegians, 2–4, 5, 9
Glaisher, James, 18, 57, 63, 97
glass tubing, 112, 137
'gun' marine barometer, *see* FitzRoy marine
Harris, William Snow, 5, 14
Herschel, Sir John, 32
Hicks, J. J., 50
Howard, Luke, 14, 18
Hughes, Henry, 74, 76
Hydrographical Office, 6, 15, 18, 40
Jones, Thomas, 6, 12
Kew Observatory, 42, 50
Kew pattern marine barometer, 46, 47, 50, 51, 52

King, Captain Philip Parker, 4–5, 11
Kirk, T., and Co. Ltd, 77
The Life-boat (RNLI), 62, 63, 64
lightning conductor, 5, 14
lunisolar theory, 32
Magnetic and Meteorological Department, Greenwich, 18, 57
Marié-Davy, E. H., 32, 35
marine barometer, 6, 12, 19, 40–5, 122–6; *see also* fisherman's aneroid; FitzRoy marine; Kew pattern marine
Marratt and Ellis, 112, 115
Maury, Captain Matthew Fontaine, 19, 37
Mercantile Marine Boards, 16
mercury tube, 130–2, 136–7
Meteorological Office, 19–21, 29, 30, 35, 39, 40, 54, 61, 67, 84, 123, 127
mountain barometer, 120
Narrative of the Voyage of HMS Beagle (FitzRoy), 3, 6, 14, 16
National Physical Laboratory (NPL), 123, 124
Negretti and Zambra, 42–3, 88, 90–2, 118; aneroids, 78, 79–83; marine barometers, 45–8, 51, 52, 53; storm barometers, 54–5, 57–8, 61–2, 63, 67–73, 74, 127, 128
Newman, J., 6, 125
Northumberland, Algernon Percy, 4th Duke of, 63
plugging device, 92, 93
pocket barometer, 118, 121
Reid, Lieutenant-Colonel William, 18, 19
Royal Charter storm, 22–3, 24
Royal National Lifeboat Institution, 39, 58, 59, 62–4, 65, 74–9

Royal Polytechnic barometer, 101, 108, 111, 112, 113
scale words, 55, 56, 85, 116–19; *see also* weather rules/words
scales, 48, 50, 55, 123–5; paper, 84–5, 101, 132–5; porcelain, 55–7, 129, 130
sea coast barometer, *see* storm barometer
Shipwrecked Fishermen and Mariners' Royal Benevolent Society, 82, 83
Simmons, George Alexander, 92
sliding level indicators, 137, 139
Stebbing, George James, 9, 64, 66
storm barometer, 58, 59–62, 68, 69–74, 75; buying, 126–32; design, 54–5, 57–8, 63; distribution, 64–7; plates, 59, 64, 65; RNLI, 62–4; scales, 55–7, 84
storm glass, 88–92, 93, 136
storm-warning system, 23–9, 34, 111
Symons's Barometer Table, 85, 87
sympiesometer, 9, 11–13
synoptic chart, 23, 24
thermometer, 69, 70, 71, 129, 132, 136
The Times, 23, 26, 30–2
transport device, 104, 105
Treatise on Meteorological Instruments (Negretti and Zambra), 45, 46–8, 118
vernier, 123, 129–30
Victoria, Queen, 29
The Weather Book (FitzRoy), 24, 25, 27, 32–5, 36, 37, 84, 85, 89–90
weather chart, 23, 24–5, 26
weather forecast, 29–32
weather rules/words, 74, 84, 112, 115; *see also* scale words
wooden case, 61, 128–9, 134